THE NEW DIGITAL WORLD ORDER

A GUIDE TO HOW GOVERNMENTS AND BUSINESSES CAN BUILD BACK BETTER IN THE NEW NORMAL

Khurram Hamid

FIRST PRINTING, January 2022.
Harry Markos, Director.

Paperback: ISBN 978-1-914926-68-6
eBook: ISBN 978-1-914926-69-3

Book design by: Ian Sharman

www.markosia.com

First Edition

I'm thankful to God for giving me the strength and life to write this book.

To my mum, Fozia; dad, Hamid; and sister, Amna, for giving me the best childhood, full of warm and loving memories, and for always being there for me.

My wife, Nagina, my rock, my love, my everything.

My three kids, Aiza, Anaya and Ayan, who drive me to ensure in some way I can create a better world for them.

Chapters

Preface

I had been diagnosed with severe COVID pneumonia and was in the ICU. Fifty percent of my lungs had COVID-19, and I was told I was in a critical condition. It was early February 2021, and both my wife and I had contracted COVID-19. However, on my seventh day I developed breathing difficulties and was rushed to the American Hospital in Dubai. On my second day in the hospital, the doctor came to me and told me that my life was in grave danger and my situation could get worse fast. They would try a new therapy, a mixture of steroids and an antiviral therapy called Remdesivir for five days to bring me back from the brink. It would be ten years to the day that I started my fight with cancer and won. Now again it was time to find that inner strength, that belief and that hope from God, and channel all the positive thoughts and positive energy from my loved ones to fight again for my life. As lay in the ICU fighting for my life, something happened. I had a moment of extreme clarity, a moment which gave me the strength to fight and to focus. I have had the privilege to be a pioneer over the last twenty years in my field, driving digital transformation and future foresight for industry powerhouses such as Gillette, Procter and Gamble, GlaxoSmithKline, and Pfizer. I had been the change agent, the intrapreneur always breaking new ground and evangelising digital technologies and their impact on consumers and customers. I enjoyed pushing the envelope of innovation and developing strategies that were disruptive and broke new ground. I had the opportunity to devise and consolidate that thinking when I helped developed the digital marketing curriculum for the University of Cambridge Judge Business School MBA program.

Writing a book was on my bucket list, inspired by my late grandfather, who was a pioneering journalist from South Asia. He was known as the King of Headlines and had written a book called *Stop Press: A Life in Journalism*. From a young age he was my inspiration and role model. He had always broken new ground and challenged the status quo with his words. I did not realise that my time in the ICU and my brush with death would give me that moment of clarity and focus to write this book. I have always been very aware of my surroundings; I was very lucky thanks to my parents from a young age to travel the world and experience different cultures, and my natural inquisitive nature and keen interest in politics from the age of nine, I would say was down to my grandfather being a journalist. There would always be debates on politics with him and my uncles over the dinner table at their home. I now believe that environment at that age gave me self-awareness and understanding as well as interest and tolerance to respect different viewpoints but also to have the confidence to speak my mind.

On my third day in the ICU, my condition completely turned around and improved considerably. The therapy which the doctors had prescribed had worked, but there was still a long way to go. This sense of euphoria gave me the idea to write this book. The title came to me at around two in the morning while lying on my hospital bed. The narrative was clear: the world was in the grips of a world war, but the enemy was invisible, and what we were witnessing was a biological holocaust on our social media feeds. After every world war, there is an opportunity to build back better, socially, economically, politically, environmentally and culturally.

Digital was transforming every part our life before the pandemic; however, the pandemic accelerated that. It made decisions for business and government leaders to digitize because they had no choice but to. COVID-19 was the best chief digital officer the world has ever seen.

This book combines the opportunity of post-pandemic reconstruction and the impact of further digitization of humanity. It provides thought-provoking ideas and insights to the limitless possibilities for post-pandemic reconstruction from a social, economic, political, enviornmental and cultural perspective, presenting a blueprint and guide to anyone who wants to build back a better society and a world which is fairer, more just, happier, and more prosperous.

Whether you are a CEO, government official, or student, the idea of this book is to help provide a blueprint on how to leverage digitization and make you aware of the key trends which may impact various aspects of humanity. A lot that is written in this book is possible today. All that is required is the will to build back a better roadmap which challenges traditional behaviours and structures to adjust to the new normal. The pandemic showed how interlinked we are as a human collective. We literally live in a global village, from a market in Wuhan to a biologicalholocaust. We need to think differently as leaders and use this opportunity to have a global reset. This book has been designed around starting that debate, giving a base of thinking to build from, ideas which can be localised and implemented, pushing the boundaries of disruption and creating a new digital world order. A world which is better than yesterday requires improving what we have today and building a better tomorrow. This book does not discuss how COVID-19 happened. However, I have used COVID-19 as the watershed moment, the disruptor, and the chaos candidate that will lead to changes to how we live on earth forever. It does try to focus on the positive opportunities that such a disaster has brought humanity, providing a unique insight into what is possible. This includes how we can own and monetize our own data and how governments need to act and behave differently with the use of data and digital technologies and the opportunities they bring to create new digital economies and jobs. It also includes the challenges digitization will bring, the resistance to change that will come from every corner of society due to the impact it will have on society. Some of the ideas in this book are very disruptive, really challenging traditional thinking, norms, and ways of doing things. As we build back better, the purpose of this book describes where we are, where we can go, and the potential challenges that can occur. These are my viewpoints and ideas, having been a digital thought leader for Fortune 500 companies and G20 nations and having the knowledge and experience of being in the middle of the disruption of technology, business strategy, and public policy.

I want the reader to take the essence of these ideas and use them as a catalyst to drive change which makes sense for the situation, identifying opportunities that can be executed and not just remain good ideas that are hard to implement.

Some of these ideas can be executed now, some may already be happening, some may take a generation to execute, and some may just be too farfetched and disruptive to ever occur.

The world is going through a major disruption on many fronts. If this book can help make that disruption more manageable and help businesses and governments build back better and be more proactive rather than reactive to manage the chaos of the new digital world order in front of us, then that would be success.

Over the remaining days in the hospital, I had written the titles of ten chapters of this book, and within the first month of being back out of the hospital, I had written the first fifteen chapters fully. For me, this book was my therapy. It helped me fight my personal war with COVID-19 by giving me a sense of hope. During my recovery at home, it allowed me to channel my energy and time to do something productive, something that would leave a legacy, an imprint, and a gift for the world just like my grandfather did when he wrote his book.

I hope you the reader have "aha!" moments while reading this book and that it helps you, potentially triggering constructive ideas and solutions to problems which exist today or will emerge tomorrow. History will judge all of us for what we did in this world, how we contributed to making the world a better place for future generations.

At a time when the world requires thought leaders, mission-driven critical thinkers can help evangelize a new way forward that is better, fairer, safer, healthier, and more prosperous than where we were yesterday and where we are today—the opportunity to build back better.

Chapter 1
Build Back Better

A pandemic is a biological war with an invisible enemy. We are living through World War 3, where our enemy is a disease which cannot be seen and our military and intelligence apparatus is redundant.

What we are witnessing is a biological holocaust, within our social media feeds, with millions if not billions of people infected and dead. What is most surprising is the hardest hit nations are not the developing nations; it is the rich Western nations, the nations which seemed to have the best health care system and infrastructure in the world. These Western democracies with all their money, power, intelligence, and technology are witnessing a mass culling of their own populations. COVID-19 is killing and infecting sicker, older, and poorer people, including ethnic minorities. In the name of herd immunity, the baby boomer generation is being wiped out throughout the Western world.

We are also witnessing the end of the American Empire, just like World War 2 brought an end to the British Empire. Due to the pandemic, we now see the implosion of America from within, culminating in the attack on the Capitol Building and the severe divide between the right and left along with the emergence of fringe right-wing groups gaining a mainstream voice.

What has strengthened during the pandemic war is our over-reliance on digital technologies to keep connected, increasing our addiction to our mobile phones and the constant need for social presence and appreciation. Consequently, the Silicon Valley technocrats have consolidated a massive amount of power.

As we come out of this pandemic war and start to live our lives and function as a society in this new normal, things have changed. Just like after every world war the world changed—the political, economic, and social system changed—society changed, it progressed, it learnt from what had happened. We now must move ahead to the post-reconstruction of the global model of society, where we all live in a hyper-connected cloud. The cloud unites us, algorithms drive us forward. We are no longer tribes, or even citizens of a state. We are citizens of the cloud living in the Metaverse. Without the cloud, we have nothing. With the cloud we believe we have what we need to live and thrive.

The use of digital technologies intertwined in a digital physical environment will form the basis of the new digital world order, which will form the basis of new economic, social, and political movements and thinking.

The scars of COVID-19 will last for generations. It will be the new mission thinkers that will drive this new digital world order to a happier, more peaceful, equal, and just union between people, technology, the environment, and the land.

To build back better, there will be massive changes to how we operate socially, economically, and politically in this new normal. These changes will not be immediate, nor will they be gradual. There will be confrontation, protest in the name of freedom, liberty, and privacy. Technology has taken over every aspect of our lives. We have given away a lot of privacy in the name of access to free internet services such as social media. The new digital world order requires a reset, so we can build back better, a just social contract between people, governments, and businesses to create a safer, happier, more sustainable, and more equal and just global society.

The pandemic war has exposed the fault lines between global collaboration, and collaboration with countries between central and local governments. Especially in Western society, there are massive fault lines between the rich and poor, with this divide getting bigger and bigger. It is a ticking time bomb where you may see a Western Winter, spring, or autumn like the Arab Spring ten years ago. We can see that through the rise of White Supremacists, The Woke, and the Black Lives Matter movements. After every world war, we see the rise of powers and the fall. In the Second World War it was the rise of the United States, as it was unscathed from the scars of war, unlike the European empires who had to rebuild. We will again see this happening. Whichever country can come back to normal the fastest after the pandemic and rebuild the fastest will play a key role in post-pandemic reconstruction.

Globalisation brought people together, creating global cultural icons, celebrities, brands, products, and services with uniformity across borders. A Big Mac is a Big Mac anywhere in the world. It looks and tastes the same, with minor localisations. "Go local," became the new mantra as multinationals expanded on a global scale with a hint of localisation, based on language, taste, or culture. E.g., you would never find pork on the menu in a McDonald's in a Muslim country or beef in India. With the advent of the internet and the language of the internet being in English by default, the Anglo-Judaic culture and its social, economic, and technological belief systems became the overarching global standard of globalisation. Accessing the internet and using services such as Facebook, required the creation of a new language, one using a combination of images called *emojis* and *memes*, to use the English alphabet to communicate in Arabic, Urdu, Hindi, and Mandarin.

However, globalisation for the last ten years has started to unravel, with the rise of the alt-right and nationalism, especially in Western society, which culminated in Brexit and the election of Donald Trump. These two key events were a signal to the Anglo-Judaic political elite that the people were not happy with the direction of globalisation. Globalisation may have lifted billions of people out of poverty and increased and expanded the middle class in developing countries, but the benefits of globalisation in Western countries was not being felt. Instead, jobs were moving abroad, factories were being closed, and the queues at food banks were getting longer. Austerity had created deep divisions in Western society and a lack of trust in government and the political class. The COVID-19 pandemic exposed this. People didn't believe or follow their governments when it came to wearing masks or social distancing. The anti-vaxer movement brought rampant conspiracy theories on COVID vaccines, and track-and-trace apps were not being followed or downloaded. Therefore, the spread of the virus in Western countries far outpaced that of anywhere else. It became a question of freedom, civil liberties, and human rights, instead of common sense and human safety. It was further exacerbated with misdirection and mismanagement of the pandemic, especially in the UK and the US, where the pandemic hit the hardest. In the UK you had the prime minister's key advisor breaking lockdown rules and a president in the United States who played down the importance of protection against COVID by never wearing a mask. This further exposed the divide: one rule for the elite and one rule for the rest. As we emerge from this pandemic world war, the distrust and the division which already existed before the pandemic has been further exacerbated because of it and as we move into post-pandemic reconstruction will be the reason for a Western Winter or Spring once the sugar high of post-lockdown freedom ends.

Chapter 2
The Great Reset

Communism lost and democracy won; however, it is China, a communist nation that is now the global superpower. The Western-style democratic model of governance, the one man, one vote system, combined with free-market economics post the Second World War led to the biggest rise in economic and social prosperity and the biggest decline in poverty the world has ever seen. The balance between social welfare and free-market economics with the goal of uplifting and providing a social safety net for the most vulnerable, executed after the Second World War by mission-driven thinkers and leaders, created a model that was worthy of export to other countries because it worked.

But since the start of the twenty-first century, this model has started to unfold on itself, through greed, corruption, and hubris. The advent of neo-liberal economics blurred the lines between a government for the people and by the people and a government driven by big business and lobbyists. Big businesses avoided tax but were the biggest recipients of taxpayer money when they were close to collapse. Plus, western governments spent more money on funding foreign wars and bailing out banks, rather than upgrading crumbling infrastructure.

Ever since the financial crisis of 2008, Western governments have lost direction, combined with the rise of new emerging economies with highly skilled and young populations such as India, China, UAE, Singapore, and others.

The Western political elite lost sight of what made post–Second World War a golden era for their development. However, throughout human history, great civilisations such as the Romans the Muslims all had their golden periods of enlightenment which came to an end. We have had the opportunity to live through and be part of the Anglo-Judaic period of enlightenment.

The contributions of this golden period and the achievements made will form the fundamentals of the next phase of the post–COVID-19 world, which will be based on fundamental and universal truths that have endured as part of human existence for centuries.

The basis for every country in the world is a constitution, whether that is codified or non-codified.

A constitution guarantees some fundamental rights of citizens and the relationship between a state and its citizens, including the social contract.

This document is very old. It is amended, but still is very old, for some countries hundreds of years old.

It has been broken and amended by leaders throughout the years to justify martial law, rigged elections, the war on drugs, the war on terrorism, etc.

In the post-pandemic world where we build back better, why do we need individual state constitutions? Why can't we have one for the world with certain localisations that factor in cultural and religious sensitivities? Is it possible to allow people to submit and vote on what they want in a local constitution? Would it be that difficult to define the rights and freedoms of humanity in the 4th industrial revolution?

The Great Reset is about looking at all aspects of what was wrong and what was right before the pandemic war and to further improve and modernise it so that is relevant for an interconnected society that exists in the 4th industrial revolution.

These changes will take time and will require major adjustments. They may seem radical, but radical thinking is required, as we have a big opportunity not seen since the Second World War to build back better.

Is democracy the only means of representation? China has been successful in uplifting a billion people out of poverty through a centralised communist system. The monarchies in the Middle East, especially the UAE and now Saudi Arabia, are accelerating development and are developing world-class cities, with world-class infrastructure, true marvels of the modern world with robust social welfare systems that ensure their people have a social safety net. One size does not fit all, and there is no perfect system. The global reset is not about creating the best system or which system is better. Any system of governance needs to focus on ensuring health, prosperity, harmony, and happiness amongst the people and the environment. This allows people to bring up their families with financial security within an equitable and just society that is respectful of people's viewpoints and tolerant of people's differences, where one life is no different from another and everyone is equal. The Great Reset should also be seen as a great equaliser of society. Due to the gap between the rich and poor, the haves and have nots, the elite and the rest, that discourse has arisen, and instability and chaos can arise. We are witnessing a perfect storm where the impact of new digital technologies together with the pandemic war and the distrust in government, the globalisation of society, and climate change have created the opportunity for the Great Reset for the world. What has been successful over the last thirty years is globalisation with localisation, or *glocalisation*. Having a glocalised constitution which is crowd sourced online, not just for governance or the rights of individuals but also the rights of businesses and how they should operate, should be included. Post-pandemic reconstruction will require the reset of this fundamental component of how modern society is organised and operates within the new digital world order. As we build back better, current and future leaders in government, in business, or in civil society in general should not see a glocalised crowd-sourced constitution as a threat but as way to seek the wisdom and viewpoints from all its citizens, so they feel like they are part of post-pandemic reconstruction, that their voices are heard and trust is built between citizens and government. Such windows of opportunity rarely occur. It will be those smart world leaders who have the will and want to build back better who will see such an initiative as way for their countries to progress and build a more inclusive society. This approach can be applied to any type of governance model, as a constitution is fundamental for any country to organise itself, whether it is a democracy, a monarchy, or even a dictatorship. Throughout the last thirty years we have seen democracies that behave like dictatorships and monarchies that behave like democracies. The use of digital tools to streamline access to government services has already started. However, the use of such tools to address fundamentals of how government behaves and operates has not happened. The new digital world order, in essence, is built on using such digital tools to build that more inclusive governance model underpinned with a modern constitution designed for the people by the people.

Chapter 3
The Rise of the 5th Estate

Before the French revolution, French society was divided into three orders or estates of the realm – the first estate (clergy), second estate (nobility), and third estate (commoners). With around twenty-seven million people or 98 per cent of the population, the third estate was by far the largest of the three—but it was politically invisible and wielded little or no influence on the government. As might be expected in such a sizeable group, the third estate boasted considerable diversity. There were many different classes and levels of wealth; different professions and ideas; rural, provincial, and urban residents alike.

Members of the third estate ranged from lowly beggars and struggling peasants to urban artisans and labourers; from the shopkeepers and commercial middle classes to the nation's wealthiest merchants and capitalists.

Despite the third estate's enormous size and economic importance, it played almost no role in the government or decision-making. The frustrations, grievances, and sufferings of the third estate became pivotal causes of the French Revolution. Not all members of the third estate were impoverished. At the apex of the third estate's social hierarchy was the *bourgeoisie* or *capitalist* middle classes.

The bourgeoisie were business owners and professionals with enough wealth to live comfortably. As with the peasantry, there was also diversity within their ranks.

The so-called *petit bourgeoisie* ('petty' or 'small bourgeoisie') were small-scale traders, landlords, shopkeepers, and managers. The *haute bourgeoisie* ('high bourgeoisie') were wealthy merchants and traders, colonial landholders, industrialists, bankers and financiers, tax farmers, and trained professionals such as doctors and lawyers.

The concepts of an estate, which represents parts of society, still exists today. Another estate grew in power, and that was the fourth estate. The term *fourth estate* or *fourth power* refers to the press and news media, both in explicit capacity of advocacy and implicit ability to frame political issues. Though it is not formally recognized as a part of a political system, it wields significant indirect social influence.

The fourth estate has wielded considerable power over every aspect of society. Starting first with print then radio and then television, creating a check and balance between the political and business class. Control over the media has always been a critical component for governments to control the narrative and influence their populations. Whenever we have seen coups in any country, control over TV has been critical to control the narrative and show power. What the eyes see, the mind believes, and television media has been used as the key tool for propaganda. The media has evolved considerably. Television media started off as state-backed television channels. This become the main voice piece of the government of the day; headlines were controlled. Economic liberalisation in the '90s led to private channels appearing and then consolidation with media barons such as Rupert Murdoch, who controlled a massive media empire across countries. In the UK during an election, whoever the Murdoch press was supporting would probably win the election such was the power of the press media.

The advent of the internet led to a massive disruption of how media was consumed. Suddenly, anyone with an internet connection could be a publisher; anyone with a mobile could become a producer. A single person with a YouTube channel could generate more views than a mainstream newspaper or TV channel. In the last 10 years people are now buying fewer newspapers, watching less of mainstream TV channels, and consuming news on demand rather than waiting for the nine o'clock news. Bloggers are carrying more influence than trained journalists, and a single tweet, even fake, can start a riot. Fake news began, and the massive social media platforms such as Facebook and Twitter became the new news platforms where people got their news. Even this was based on an algorithm, censorship became personalisation controlled by an AI.

The fifth estate really became prominent during the election of Donald Trump as president of the United States of America. The use of digital marketing, where data was used to target ads based on online behaviour, was used as an election tool. Online ads, which had been used to convince people to buy stuff, were used to promote political headlines or clickbait. Behavioural data gleaned through platforms like Facebook were used to personalise political ads which pointed to legitimate-looking fake news sites run by countries looking to destabilise another country using psychological warfare.

Consequently, it became close to impossible to control the media. Even the traditional fourth estate media was reporting on news from social media.

Social media has become mainstream media, and the fifth estate have become a powerful global estate which gives anyone with an internet connection the opportunity to create and produce content. Social media millionaires have emerged, from seven-year-olds opening toys and getting millions of views to people watching people playing video games. This major shift in content production and distribution has created a period of flux. The previous traditional four estates had created an understanding and an equilibrium with each other, which has now been totally disrupted by the fifth estate. We now have social media influencers who have become the new news reporters, gaining millions of millions of followers. Clickbait-based headlines has become the normal way to get viewers to gain advertising revenues. Disinformation through social media, which can create chaos, is now the medium of choice. Algorithms and artificial intelligence are personalising news based on an individual's behaviour.

As we build back better, this fifth estate and the social media influencers who are the new stars of the medium will need to be regulated and held accountable. The fifth estate can no longer exist as part of the state, which does not have accountability or transparency. It now wields too much power. The fifth estate should be able to challenge the status quo and hold other parts of the state accountable, from government to business, but it also needs to act responsibly, so it does not create chaos, social discord, resentment, and fear. The creation of a citizen-based accountability board, an independent rating system to rate content for its factual accuracy and a board which monitors and performs the role of a regulator to ensure the independence of the fifth estate from any political bias and influence but ensures transparency and accountability, will be a critical component of the new digital world order as social media has become mainstream media.

The regulation of the fifth estate will require a global approach. As social media platforms are global, a global approach to their regulation and accountability will require cooperation from all countries.

All governments are trying similar approaches to regulate social media, data protection and privacy, and new rules around hate speech, online bullying, etc. Sometimes in some extreme cases, governments block access to these social media sites.

Social media companies and governments have become frenemies. On the one hand you have leaders using social media to communicate to their citizens, and on the other hand social media companies, as was the case with Donald Trump, can shut an account. Anti-vaxers are being shut down during the pandemic to stop spreading misinformation. The question remains how to balance freedom of speech with responsibility. This will be a key issue in the new digital world order. Information and news always have a bias, based on the individual's belief and interpretation. Social media, due to its global nature and the ability to upload content, is being used for disinformation. However, disinformation is also an opinion and a viewpoint. The problem with social media is viewpoints are expressed by deep fakes which are sophisticated AI impersonating a human being which can cause civil unrest due to the fact state or non-state actors are taking advantage of this type of technology to sow social discord. Unlike broadcast or print media, which has always been one-way, social media has allowed for a two-way dialogue and the freedom to express viewpoints no matter how extreme or liberal they may be, along with the ability to share them with a global audience.

To qualify content on social media, some type of fact-checking would need to be created through AI, which is unbiased. This cannot be a fact-checker that is developed or owned by social media companies. It would need to be done by an independently funded company that doesn't have any financial or social gain in playing the monitor and checking facts. It can also use the wisdom of the crowd and independent fact-checkers from the countries from where the news is coming from. The idea of having this independent AI and human-driven fact-checking system is not to stop content but to provide the reader data points to show if the news is accurate and correct or biased. This would be for all news content on social media, including global and national organizations such as the BBC, CNN, etc. It would improve the ability to highlight fake news, disinformation, etc., faster than what exists today.

Chapter 4
The New Superpowers

The 4th industrial revolution, which really accelerated in the last five years, has shown the best of the human mind. Advances in technologies, innovation, and creativity have also shown the worst of technology and the worst of humanity. The internet has brought us closer together in real time but has further polarised us through algorithms. These algorithms decide what we should watch and consume based on our online history and the behaviour of our social network. This basis of censorship is unprecedented. Every great empire and superpower have tried to conquer the full spectrum, the main one being the battle for hearts and minds through manipulation.

Controlling the narrative, messaging, and the press, which *was* done by governments, is now done by private tech companies who have developed algorithms that have content moderators, and anyone who posts content that is against their terms of service is censored.

When the leader of the United States, the so-called superpower, the commander and chief, the person who has access to the nuclear codes of the most powerful military in the world can be banned from Twitter and Facebook with a click of a button, we must all realise who are the new superpowers of the 4th industrial revolution.

But such power over billions and billions of people who live every moment of every day using these platforms, with no oversight from any independent body, is not good for any society. Our global society is now intertwined with the use of these platforms. They have provided us access to information and connection to our family, colleagues, and friends, especially during the COVID-19 lockdowns. They have given us unprecedented access to new markets, new knowledge, and empowerment. These Big Tech companies have contributed so much to the prosperity of the human collective. They are now too big, too monopolistic, and global oversight is required. I don't mean oversight in the traditional sense by the old, haggard bureaucracy, but oversight by the users of the platform, who become the moderators, the independent regulators not connected to any government or political party.

The empires of Facebook, Alibaba, Apple, Google, Amazon, have unprecedented control of multiple aspects of society, from banking, health, and media, to advertising and transportation, etc. However, this is not country specific. This is across borders and direct to our screens on a very personal, unique level.

We are living in a time where we are seeing the end of the power of the nation-state, as control over populations, economics, the media, and society in general is consolidated with Big Tech. In the new digital world order, these companies would need to be broken up. As data becomes the new oil that drives the 4th industrial revolution and underpins the new digital world order, we must ensure that we don't develop data colonies. These are countries and citizens who are used for their data but don't gain anything in return and essentially are run by private tech companies. If we look back in history, we can refer to the Dutch East India Company as an example of a company that was used to expand and colonise the world. The modern examples of this would be Big Tech.

As discussed previously, Big Tech needs to be regulated by its users rather than a government body, and governments need to adjust and apply the technologies used by Big Tech to build a better representative model, a more efficient governance system, and a more inclusive society where people represent themselves. The expertise and technology of Big Tech can be used for this. Build back better will require Big Tech to play a greater role in the rebuilding of a post-pandemic society in partnership with governments with citizen oversight.

It will also require striking the right balance, where Big Tech companies can innovate and create new services and new tools but ensuring that the influence and power of these companies are kept in check to ensure that innovation and creativity are not stifled.

As these companies have contributed so much to humanity in the last twenty years and are still trusted by their users even more than governments, the breakup of these Big Tech companies should not be taking a hammer and breaking them up. Instead, it should be more about their operating models and decision-making processes, especially when it comes to ensuring how they use their users' data is different from it is today. Regulation and oversight should not be overkill, nor a light touch, as self-policing does not work. The use of digital platforms is strategically too important to remain the way it is; a new model is required which strikes the right balance between innovation, creativity, efficiency, experience, trust, and profit. User governance boards should form a core component of the operating model of these tech companies; however, these user boards should be elected by the users of the platform directly and should exist in each country the platforms operate in. Creating the right checks and balances is important to avoid the consolidation of too much power and control in the hands of a few companies who have billion and billions of users and know everything about those individuals twenty-four hours a day, seven days a week. However, we must also appreciate why billions and billions of people have adopted these platforms. What are the reasons? Is it that they are designed to provide an easy user interface and experience which allows a broad spectrum of users to be addicted to these platforms and have the confidence to share every aspect of their lives every day voluntarily, from where they are, to what they eat, to what there thinking, with the only reward being a like, comment, or share? In the traditional sense, these platforms have become the modern-day public square, a meeting point, a living room where people interact and communicate, but now it can be done with anyone anywhere in the world in real time. What has not occurred is figuring out how to leverage or create similar platforms that can help in all aspects of post-pandemic reconstruction using the wisdom of the citizen.

The tech race has now become like an arms race. Whoever can develop the most sophisticated AI, the fastest-processing supercomputer, or the smartest engineers and cyber specialists will be able to advance and progress in their development and protect themselves. Wars will no longer be fought on battlefields with two armies looking at each other across no-man's-land, and it will no longer be with countries or even groups. Whoever has the tech strength is powerful, whether that is an individual, an organisation, or a government. The very notion of superpowers will not be about having the mightiest army in the traditional sense. It will be more about the speed and agility of a cyber-army, who is able to disrupt and disseminate quickly. That is why data combined with artificial intelligence is important to be able to proactively protect and reactively combat attacks. Keeping the public safe is the role of government. However, if there is no trust in the government and the public views government as a threat, then something needs to change. When a society rejects track-and-trace apps provided by its government to help combat the spread of a deadly virus but are more than happy to share each aspect of their lives with social media platforms in exchange for free advertising, then that society is broken. The UK is an example where this has occurred. The UK public would rather be tracked by private social media companies than a hundred-year institution like the NHS, which is focussed on public service and stopping the spread of a killer virus.

Chapter 5
Surveillance Capitalism

Each click, each share, each tap, we are in a state of permanent surveillance. It's called *surveillance capitalism*. This is when we as people and what we do online, how we behave, how we interact, is sold to the highest bidder in the form of digital advertising. It is what makes the likes of Facebook, Google, and Alibaba billions of dollars. The product is us and how we behave. Our entire online existence is packaged into audience segments and sold in real time on advertising exchanges. Think of it as the stock market, where you are the stock. We give away all this data for free so we can use these global platforms to communicate, connect, and transact. We have all done this by signing up to the terms of service of these global platforms.

However, why should these massive global platforms continue to make money from our online behaviour while we as individuals get nothing, just a bunch of ads to prompt us to buy things we don't really need? In the new digital world order, we will take back control of our data and make money from our data. Considering the impact of a post-COVID world where there will be mass unemployment and job loses, allowing people to make money from their own data becomes a source of universal income. It also allows advertisers to connect on a one-to-one basis with the customers who want to connect with them, reducing the amount of ad fraud and waste that currently exists in the digital advertising system. This ad fraud and ad waste is directly linked to funding terrorism and organised crime. At the same time, it is promoting disinformation and clickbait, which is linked to fake news, which is becoming the main source of news for so many people. This is having a destabilising effect on countries and communities.

The quantified self is a movement, a way, and a means to measure every aspect of our lives, from our health and wealth to our sentiment and feelings. With location-based mobile apps, personal fitness trackers, IOT medical devices and smart home sensors, and facial recognition technology, it's possible to measure, monitor, synthesise, and model everything about ourselves and even our family, friends, and loved ones.

Our health, location, and online behaviour combines to give a quantified version of us and our own personalised algorithm. The personalised algorithm is like our digital twin. It will help us and supports us in maintaining a balanced life, which keeps us happy and healthy, giving us a great quality of life. It will also help us monetize our data, so we maintain a certain level of basic income, social welfare income from our social media usage.

Privacy in the new digital world order will be different from before the pandemic war.

There is no privacy in the digital world. We have given that up for free access to internet services provided by Big Tech companies.

In the post-pandemic, global society, we will require a new fundamental bill of rights, which is the union between government, the people, and private tech companies. We can use the current bill of rights, which was formed in a similar case hundreds of years ago, but we need to update and bring it into the post-pandemic global society and the new digital world order. Our personal data has become a currency which we need to control and monetize, when we choose, with whom we choose. This will become a key fundamental pillar of the new universal bill of rights for a post-pandemic global society. We need to build back better and ensure we maintain, update, and upgrade fundamental rights in line with the 4th industrial revolution. We also need to ensure that these fundamental human rights, when they are codified and developed, are done so by the people, using the wisdom of the crowd to create a global bill of human rights where whoever wants to vote and contribute ideas can do so without fear or repercussions. We must ensure technology helps people, helps society, and creates happiness and prosperity for everyone.

Data has become the new oil and we are the new natural resource. Trillions of data points are generated every second of every day. People are volunteering to give away their data: where they went, what they are looking for, and what they bought. Big Tech has become the new oil companies, as data now drives the 4th industrial revolution and digital economies. Drilling for data is when companies set up in different countries to harvest that countries' data. Data harvesting and the packaging of that make internet companies money. However, the benefit of that data harvesting is not felt in the countries where that data is harvested. Whatever money is made from these data colonies is sent back to Silicon Valley or Shanghai.

In China, for example, a social rating system has been developed through the quantification of data that the individual generates. Based on this, the Chinese government can synthesise and segment individuals, giving them a rating. Based on this rating, you can either be allowed to travel or not, get preferential treatment at a government office or even a restaurant. The social rating system is a way to motivate and manipulate the population. If a citizen does something good for society by helping others, giving charity, or just driving safely, they are rewarded with a better social rating. This creates positive behaviours and motivates people to do better and be positive role models for society. The equivalent in the West would be the credit rating system. Based on how good your credit is, banks give you better offers and incentives, so essentially there is not much difference between a credit rating system and a social rating system. The only difference is one is run by private corporations to motivate you to spend more money and get into more debt, and the other is a way to create and motivate positive role models in society. People will disagree and say the China social rating system infringes on civil liberties and creates a Big Brother state. But if encouraging positive behaviours that benefit the individual and the state creates happiness, social cohesion, prosperity, and a healthy population, maybe some type of social credit system is required. Different rating systems exist already, whether it's your credit rating, the number of times someone likes, comments on, or shares a post, or the ratings or reviews of a business. The only difference with the Chinese version is it is something that was developed by the government. There is a difference between a chaotic, broken society and an autocratic one. The ideal state is something in the middle. A voluntary social rating system where citizens can decide what to share and when to the rating system in terms of their data and in exchange would receive financial rewards and recognition can be a key tool to help manage post-pandemic reconstruction. As we build back better, it will be important to motivate citizens and businesses to demonstrate new behaviours. We as individuals produce different types of data which can be loosely segmented into sets: behavioural, financial, social, spiritual, and health. The combination of these data streams generates signals and insights which can be used to manipulate an individual. It's known as data-driven marketing. Once the domain of companies who wanted to market and target the right product to you at the right time, it has now become a way to manipulate your opinion on a specific topic or how you should vote. This level of personalisation means messaging is different from the same entity on the same topic, but because of our data stream, it can be personalised based on an individual's trigger points. This is a sophisticated form of data-driven social engineering. Currently, legislation like the General Data Protection rule in Europe and similar countries, which is about allowing the individual the right not to be tracked without their permission and the right to be forgotten, does not go further than that in terms of the usage of the individual's data once they have given permission.

This is more to do with data usage rather than data privacy. There is no data privacy unless you switch off your internet connection and shut down your phone and become a nomad. Individuals are broadcasting data 24/7.

The usage of data requires agreed boundaries and rules, so it does not infringe on privacy, and it is not exploited by businesses. Governments will need to create AI and data regulators to regulate this area and develop effective policies and regulation which governs the usage of data and AI.

There is a great deal of discussion around data privacy. However, we are now in the sharing economy, where we share every aspect of our lives.

You may have a thousand friends on Facebook, some of whom you only met once in your life, but we are comfortable sharing pictures of our holidays and family moments, articles we liked, and even our opinions with total strangers. So, in the new digital world order, we must change our concept of privacy and what we are doing over social media. Amongst Generation Z who are the digital native generation, how many followers you have is a sign of popularity. More than ever before in human history, we share information about ourselves with thousands of people who are sitting everywhere and anywhere in the world, with a snap and a click. There is no one forcing us to do this. It has become part of habit to share what you're eating and where you're going. The practice of getting likes and comments for social endearment and social capital earned through a like, share, or comment has become like a drug. The rush of endorphins when we get lots of likes or the anxiety and depression when we don't is a real mental health problem. Social media addiction, where we live our entire lives through our social media feeds, is a real addiction. The addiction of gaining social recognition from our followers is now part of our global culture.

Therefore, those lines between what is private and what is not have become blurred. The boundaries of acceptability for different generations are different. For digital natives, there isn't a concept of privacy, but instead there is the concept of openness and living and sharing your entire life through a social media feed.

The privacy debate needs to be shifted too. If you are living your life online, then how do you make money from that data? Also, how can government responsibly use data to ensure they deliver better services to their citizens along with protecting them from security threats and pandemics, and how can businesses leverage data responsibly to help attract more customers and develop better, more personalised offerings if their customers desire.

Chapter 6
Generation Digital

Generation Z is the digital generation, the generation which doesn't remember life without the internet, born between 1995 and 2012.

This generation will be known as the post-pandemic war generation, where we transition from our current state to the new digital world order—an environment where those of this generation are digital natives. This is the generation that will drive the next wave of economic, political, and social movements.

Generation Z thinks and acts differently versus other generations before them. The key difference is they rely on their mobile devices, are hyper-connected, and social media presence and acceptance are key. They are content creators and have a different understanding of privacy, as they live their lives online and post everything they do on their timelines.

They will be the largest generation ever and currently make up around two billion people.

They are not brand loyal; freedom of expression is vital for them. They care more about the collective good and are more socially aware. The first internet generation is heavily reliant on digital devices, and a lot of them are developing attention deficit disorders.

Just like the baby boomers that were born out of the Second World War, this will be the generation that drives the new digital world order and the post-pandemic global society.

We will see massive generational changes, as what was important to the baby boomers was different than what was important to generation X and generation Y.

However, the influence and impact generation Z will have on the post-pandemic society will be transformative. Being the first digital natives, they will be the disruptors who are and will continue to expect changes to every aspect of business, culture, society, and government.

The transformation occurring through the combination of generation Z and the advent of digital technologies was already occurring; it has just been accelerated due to the pandemic war. Businesses that were avoiding digital transformation have needed to transform for survival. The same now goes for governments and representative models. These have yet to change, and this will be the biggest change that will help in the post-pandemic world. This is also the very generation that is straddled with massive amounts of debt from student loans, especially in the US and UK. It is the first generation where real income has declined, where a job is not guaranteed once you leave university. As consequence, they are more frugal and tend to rent rather than buy. Also, they tend to freelance and want flexibility in their work and can easily multitask. It's the first generation that is addicted to their mobile devices, documenting their lives through selfies and check-ins so that their friends know what they're doing, using different types of image filters, united through massive social, retail, and video platforms. They are live-streaming content and their lives to thousands of people across the world. They are a hyper-connected generation for whom personal space and public space have been blurred, a generation that is at ease with sharing their entire digital life with private tech companies whom they trust but have lost trust with their governments, especially in the West.

This generation is disrupting every aspect of the customer journey and experience for multiple industries. Many incumbents of those industries have found it difficult to change to a more digital-based customer experience, due to the management of these companies being of the baby boomer generation who are not digital natives. In businesses that do have digital native senior leaders, the shift to a digital-only business model has been faster. The pandemic, which has been the best chief digital officer the world has ever seen, further accelerated the move to a digital-only business model. Due to lockdowns, people spent more time at home, buying things online, streaming and playing games along with banking online. It also created and accelerated changes, for example telehealth and remote diagnostics became the new starting point for the patient journey, as people could not go to the doctor due to the overburden of the healthcare system and the fear of getting COVID-19 from visiting the doctors. Due to lockdowns, food, grocery, and everything else was ordered online and delivered. Touchless and contactless payments became the norm, as paper money was discouraged. The start-up ecosystem globally grew at unprecedented speed and scale. It wasn't just start-ups in Silicon Valley raising funds; it was start-ups from Karachi to Cairo. All these were led by generation Z founders all focusing on solving problems for their respective countries whose size of generation Z populations together with access to high-speed mobile 3G and 4G internet is exploding and therefore creating a customer base of people hungry for new digitally driven customer experiences.

Chapter 7
Data Sovereignty

Our identity is made up of so many different data points. We have our national ID, our social media ID, our banking ID, our employee ID, our medical ID, our credit rating, etc.

A national digital identity which connects to our citizenship or residency will be a key building block of post-pandemic reconstruction and the construction of a digital economy. We already have our Facebook, Google, and other IDs which are global and allow access to multiple different online platforms. These IDs are borderless.

The concept of self-sovereign identity is based on using Blockchain technology and having different stakeholders validate an ID, so that it can be a combination of your online Facebook ID, national citizen ID, medical ID, credit rating, and employee ID. These nodes act as validators, which means you can essentially produce a global digital ID system. The benefit of a global digital ID system that transcends borders allows for more efficient movement of labour and business across national borders. Suddenly, with a global self-sovereign identity system, we become global citizens. We become one massive global community, we become empowered, and we are therefore not dedicated to a specific state, where we live. We will obviously be connected, and our self-sovereign ID will be used. But the efficiency of having a global identity system in the new post-pandemic global society becomes the basis for so much innovation and creativity of thinking to shape a new global structure that creates efficiency and scale. Our self-sovereign identity is also our own. We control it, who we share it with, and who can store it, therefore giving the individual full empowerment. Because it's based on Blockchain technology, it is immutable and therefore less susceptible to fraud.

Self-sovereign identity is part of the quantified self and the personalised algorithm. These become the fundamental components of the new digital world order and the goal of a cloud-based global society. It also ensures that the control over your data and identity is in the hands of the individual, and they decide what can be shared with whom.

Personal data sovereignty will be a fundamental human right in the new digital world order, and ownership of an individual's data will be with the individual.

However, personal data will also be required for monitoring potential new pandemics and epidemics.

Data is the new oil, so data sovereignty especially with a national digital ID system, is an important step in establishing a digital economy. The fundamental component of a digital economy is data, smart, relevant data that starts with a sovereign ID. You need a sovereign ID to open a bank account, take out a loan, or open a company. However, right now we use paper-based or card-based IDs which need to be scanned and uploaded. These require further verification with national data ID systems, which are not open and, in most cases, don't allow other companies to connect to them.

It is therefore important that governments create national open data systems which allow businesses to tap in to gain access to valid ID information. This can be either for individuals or business, fast, and in real time KYC (know your customer) verification, which is paperless and a combination of not just biometric identity but also credit or social rating.

The sovereignty of data is the key element of a data economy. Currently, data is only monetized by private tech companies. However, in the future, just like in the past, governments will play two key roles in the data economy. Data is the new oil, so we only need to go back to the middle of the twentieth century where oil companies were state-owned entities because oil was seen as a national strategic asset which drove the economy. We can draw the same parallels to data, as data now is a strategic asset which will now drive the digital economy. Therefore, governments will need to create state-backed data companies, creating open data platforms across industries anyone can API into to create new digital products and services. Financial data, health data, ownership data, social rating data, any identifiable data about individuals and businesses could be monetized by these state-backed data companies. However, the owners of these state-backed data companies will not just be the state. They will be the market, the customer, consumers, and businesses that operate in that specific sector. This would create data quality, which would enable analysis and insight gathering, which can identify problems and develop solutions. These state-backed data companies owned by government and market participants would need to incentivize citizens through payments to use their data so that everyone can benefit from the monetization of data.

Allowing state-backed data companies to be co-owned by government and market participants, will ensure data uniformity and equalise data access. Allowing start-ups to then API into the state's data assets to develop new products and services creates jobs, promotes investment, and reduces barrier to entry. The concept of *govtech*, where government leverages technology to automate government services and processes, has been around for a while under the banner of e-government. However, by creating tech and data-led state-backed entities, governments are not just accelerating their digital economy, but also they are allowing such companies to develop tech innovations that can be used by government faster. In this scenario, the government becomes on one hand an investor and on the other hand a customer of the products and services. This accelerates the use of tech in government beyond an e-visa to more meaningful changes such as using the wisdom of the crowd to develop and vote on legislation using AI and automating large strands of government bureaucracy.

The creation of a self-sovereign digital identity which is interoperable across borders through a unified global digital identity system will remove levels of archaic paper-based practices. The use of digital attestation of documentation will speed up the processes of job hiring and immigration. The cross-sharing of identity data across borders will help in solving crime and capturing fugitives. The use of tracking will allow for the faster tracing of diseases such as COVID-19 and the reduction of its spread. There will be hesitancy, especially in countries where trust between the government and its citizens is low, due to COVID-19. We will now require vaccination passports to travel or enter public places. Without this you cannot travel or go to a mall. But having a vaccination passport to travel is nothing new. To travel to many countries in the past, you needed to be vaccinated against tropical diseases such as yellow fever etc., so showing you have a COVID vaccination for traveling is therefore the same. However, scanning a QR code which shows your vaccine pass every time you enter and exit a public area is a new phenomenon because of COVID-19. It can also be seen as normal as a check-in on social media when you go somewhere. The only difference is that you must be vaccinated to enter.

With this, governments are getting into the tech space, but unfortunately, without having a unified global digital identity system where data is interoperable across borders, there will always be problems, as standards are not the same and data is not connected. This is not just the case between countries. This is the case between local, state, and national governments and different ministries. In some cases, instead of having a single COVID application for a whole country, there are several, depending on which local or state government you are part of. Then, even being in the same country, they don't talk to each other. This will be a major challenge for governments who are the issuers and owners of state-based identity: to have better and more robust solutions in place and develop a unified digital identity system which connects the whole digital and data ecosystem of the state as well as internationally.

However, one of the weaknesses of this is the reliance on app stores of big tech companies. As governments move into the tech space they are still limited, as they lack the control over the app stores of their countries. For example, vaccine passports apps are being rolled out by governments globally and citizens must show their vaccination status to enter public places. However, what happens when Apple or Google decide to do an operating system update or refuse to allow these apps to be listed on their app stores? How do governments keep the sovereignty of their tech and the apps they develop for their citizens? Does this require a new type of app store just for government apps? would this app store be developed by each government which would also have its own national operating system? Or would it be an app store which is interoperable with all the major apps store but be governed and run by governments. The debate and direction of what will happen should be a global, one that is led by supra national bodies such as the United Nations the G20 and the WHO. As this is a global problem and therefore a global solution working with big tech companies needs to be worked out, one that gives nation states app store sovereignty without stifling innovation.

Chapter 8
The 4th Bill of Rights

The Magna Carta, or "Great Charter," signed by the King of England in 1215, was a turning point in human rights. Among them was the right of the church to be free from governmental interference and the rights of all free citizens to own and inherit property and to be protected from excessive taxes. "No free man is to be arrested, or imprisoned, or disseized, or outlawed, or exiled, or in any other way ruined, nor will we go or send against him, except by the legal judgment of his peers or by the law of the land. To no one will we sell to no one will we deny or delay, right or justice." The Magna Carta. To this day, the Magna Carta establishes the fundamentals of human rights and the rule of law. From the 1) Magna Carta to the 2) US Bill of Rights to the 3) Universal Declaration of

Human Rights, each document sets the rights of all mankind. However, these documents, some hundreds of years old, have fallen out of lockstep with the changes to humanity, the progress it has made, and the rapid acceleration of the use of digital technologies as part of our lives. The world, unfortunately, is going through a volatile moment, the pandemic war, and before that movements such as white supremacists, Islamic extremism ultra-nationalism and the ultra-left. These changes have occurred at the same time, where we have become citizens of the cloud, consuming global brands from McDonald's to Amazon, where surveillance capitalism has become the main way for digital-based companies to make money. It is now time for a new Magna Carta, one written by the people and embraced by all countries and companies. Why can't access to healthcare, social security, shelter, safety, electricity, and the internet be fundamental to all mankind considering we live in a period called the 4th industrial revolution? Why are we spending money on educating a computer to develop artificial intelligence? Why are we not spending that money on increasing education access to everyone who needs it? Billions and billions of people still don't have access to clean water and basic sanitation but can stream YouTube on their phones.

As we will build back better, we must create the 4th version of humanities Bill of Rights, one which is developed by the people for the people. The 4th

Bill of Rights is a declaration of the universal rights of mankind. It includes new areas such as online privacy and the rights of your data, the right to be forgotten or tracked and traced or the role of AI and digital humans. Legislations, and laws are never as fast as technology. However, there are points in human history, mostly after wars, where windows of opportunities are created which require changes to the past to build back better, for a greater future. The importance of a new fundamental Bill of Rights designed for the 4th industrial revolution is a fundamental step to develop a post-pandemic global society. It is important, a global bill of rights that equalises society and countries and gives fundamental rights to a human being no matter their colour, creed, religion, sex, or financial position. Post the Second World War the Anglo-Judaic period of enlightenment resulted in massive prosperity and modernization. We have modernised through technology and high-speed internet has created new digital economies. However, some of the biggest problems that have existed since the beginning of time have yet to be solved. Allowing people the basic rights of clean water, medical care, electricity, and sanitisation, this is where we have not progressed. We have been able to tackle and eradicate these problems, but not for everyone. The United Nations Sustainable Development Goals have tried to address these issues but these goals are not enforceable, they are guiding principles but not enacted into law. Through the creation of 4th Bill of Human rights where countries, companies and citizens sign up contribute to its formation, within a generation, we can eradicate these problems, by harnessing the wisdom of the crowd. This ensures that everyone who wants to feel that they have contributed can and therefore have a certain level of ownership to the creation of the 4th Bill of Rights. This Bill of Rights will be accountable and tracked, so we know which governments and business are embracing a new standard to be followed. People need to feel part of something, and the ease of voicing an opinion through a comment, like, or share creates a universal language where billions and billions can take part in designing and voting for a new 4th Bill of Rights. Unfortunately, human rights, especially women's rights are still being contested. A woman's right to an abortion is still a controversial topic. In some US states laws have been enacted where those rights are governed by the government and not the woman. This is in a country which is the leader of freedom and human rights and has tried to export that as a model to other countries. Women are only starting to gain senior roles within businesses, and diversity is now a strategy in most global organisations. However, in the US, women's sanitary pads are still seen as a luxury, and a sales tax is applied in many states even though they are a necessity for women. Post-pandemic reconstruction and the development of a new Bill of Rights will require a fundamental look into many different dimensions. The right and freedoms of individuals need to evolve based on common sense, cultural and religious sensitivities, the rapid change being driven by technology, and the global nature of society. Through creating a global platform which uses the wisdom of the crowd to develop a new Bill of Rights, the people will feel part of contributing to post-pandemic reconstruction and the betterment of society.

Chapter 9
The Meta-mocracy

We now have global platforms where we can express our views without fear, share any moment of our lives, protest, and mobilize physically and virtually for causes we believe in, all from the comfort of our mobile screen. Freedom of expression, freedom of empowerment, freedom of choice are what the Anglo-Judaic period of enlightenment has brought to the next phase of human development and the new digital world order.

In the post-pandemic global society, these global platforms can form the basis of new forms of representative models, ones that enhance existing representative systems or even change them. People don't need to elect representatives when they can represent themselves, vote directly in causes, and create laws and legislation which are executed by algorithms developed with the public interest in mind. This is a wonderful time for critical thinkers to step in and drive ahead with new models of representation.

The digital democracy or the "meta-mocracy" is a virtual place just like a parliament, but it exists in the Metaverse, where people represent themselves. They mobilize, and they help design legislation and laws. These laws are executed by algorithms. Success is measured by sentiment and happiness, like a rating and review but on laws and policies. People can participate from anywhere, and it goes down to the grass roots allowing citizens to vote. The meta-mocracy is the next generation of digital democracy, one built on the Western model but given a digital twist. This is not a radical thought: people create and take part in polls and surveys; ratings and reviews drive how and what we purchase. These technologies, however, have not been brought into the mainstream political decision-making process of governance. Such a new model would take power away from the political class that governs populations. Populations governing themselves using technology is too much of a disruptive idea. However, in the post-pandemic world, these new types of models will be required because people have lost faith in their elected leaders, elected leaders have become corrupt, and lobbyists have taken over the process of defining laws and legislation in the interest of big business. Political parties built on one hundred years of right-wing, left-wing outdated ideologies have lost touch with reality and have failed. They will no longer work in a post-pandemic global society.

The people's parliament for the people, by the people, managed by the people and artificial intelligence is the future. In this new meta-mocracy, laws, legislation, and execution of government are done through a mix of human and digital platforms. Accountability is fully transparent, with no room for closed-room old-boys-club deals. The centre of this new digital democracy is a powerful AI which is designed based on happiness and ensuring the happiness and well-being of its citizens, with no other agenda. This global digital democracy will connect communities at the most basic level and ensure decision-making is down to the grass roots, street to town, town to city, city to state, and state to country. In each aspect of government, laws will be made based on the needs, wishes, and sentiments of its citizens.

People, due to technology, are more empowered than ever before, but at the same time technology can monitor every part of our lives. Generations will react differently to changes which will occur, and this will be down to the level of comfort on things like data privacy and perceived freedoms. But this is more down to change. People don't like to change. Most people are apprehensive about changing, especially the older generation. So, in any great change, there are always people who are against and who are for it. Mostly the majority is always silent, and the minority is vocal, people who believe that their freedoms and civil liberties are being taken away. The digital democracy and its forms of representation give both the silent majority and the vocal minority a way to be part of making sure that their voice is heard. We must change the dynamic of protest; protest should not be about confrontation between people and their rulers. Whatever issues the people have, there should be an outlet where they can take part and represent themselves. This is what the digital democracy or meta-mocracy is all about. It should not be taken as a new form of global government or centralisation of power. That is not what the digital democracy is. It is a new form of decentralised governance using digital technologies such as AI, Blockchain, and the internet of things (IOT). A new type of governmental model and a new type of representative model, whether that is centralised or decentralized, will emerge, combining the best of technology and the best of the human mind.

People want their voices heard and to be able to express agreement and disagreement with their rulers. Rulers need to listen and proactively adjust if there is widespread disagreement on certain topics. Opinion polls have always been used to check the pulse on sentiment and opinion. However, with the advancement of digital tools, anyone can create a poll or a survey, and analysing sentiment from social media can provide a treasure trove of data for leaders to assess and understand the pulse of the people. However, leaders still rely on opinion polls or headlines from newspapers, which is an outdated way to gauge public opinion when social media is mainstream media. Smart leaders who can leverage AI-based sentiment analysis and scoring, real-time opinion polls, and social media analysis can develop policies and initiatives and create trust between themselves and their citizens. Channelling and analysing sentiment and public opinion will ensure leaders are not out of touch and can remain popular and maintain legitimacy.

Chapter 10
AI Government

Governments, whether they are democracies, monarchies, or dictatorships, all are driven by a state bureaucracy, the organisations that implements and develops laws and policies. It undertakes all administrative activities of the state; it can also be known as the *deep state*, as this arm of government which is not elected, runs the state.

Politicians are the front office, the deal makers, the salespeople, and the bureaucracy are the back office, if there was to be a comparison to a private sector organisation.

The success of the state is how effectively it can develop and implement legislation and policies which provide economic and social well-being to the citizens of the state.

However, these unelected officials who essentially have a job for life in government have effectively created rules and regulations just to keep relevant. This leads to inefficiencies. Also, as they control vast amounts in state budgets without any major accountability, there is a lot of wastage but also corruption. Many bureaucrats are of an older generation, especially senior bureaucrats. They are analogue leaders within a digital world. They know how to manipulate the system because they are the custodians of the system. While politicians are the ones who are elected and seen by the public and are held accountable by the public, the bureaucracy plays in the shadows. To run a more effective government operation, the use of AI technology to implement, track, and monitor legislations will be required to create transparency and accountability of bureaucrats towards the citizens of the state. In the new digital world order, citizens represent themselves and vote on legislation and policy in a fully transparent way. Bureaucrats with the use of AI technology and digital platforms would recommend and execute legislative decisions approved by citizens. In a lot of cases, the AI would recommend and execute tasks using smart contracts in which Blockchain technology would be used to ensure each task is automated and executed one after the other. A smart contract is a piece of code, essentially a digital contract programmed for inputs and outputs, ensuring the successful execution of tasks. Dubai is a great example of a government that has embraced the use of Blockchain to improve government administrative tasks, ensuring speed, efficiency, and transparency and going paperless. Through the digitization of the bureaucracy, governments go paperless, creating digitized tasks which can be automatically tracked and traced to completion supported by AI. This form of digital government operations reduces the size of government, increases efficiencies, and speeds up service delivery through automation. The automation of government administration with the use of big data and AI also can be used to develop data-driven legislation based more on facts rather than emotion and political views. Including citizens in the development and decision-making process ensures that whatever is developed and implemented is in line with the wishes of the people and not lobbyists and big business. As we build back better and develop a better form of governance, it will become imperative for governments to react fast to changes, whether those are new pandemics or creating new legislation to embrace new forms of technology, from drone technology to digital money and more.

Government data and analytics will be a core element to running more efficient data-driven operations. This will require new skills within a state bureaucracy to develop and deliver the mandate of different aspects of government. Leveraging open data and API technologies can also allow private sector companies, start-ups, and entrepreneurs to develop solutions and services which can help make government operations be more efficient. Through effectively developing new government digital services, the automation of vast swathes of government operations can be conducted by highly trained and highly skilled AI operators who can leverage open government data to deliver more efficient government operations. Reducing the layers of government operations through automation but also further decentralisation of decision-making from central to local government will ensure that the delivery of government services, the implementation of policies, and regulation are in line with the needs and wants of citizens through having citizens be part of the process. As we build back better, we will need to create an effective citizen engagement model that is not just focussed on citizens' opinions and views on rules and legislation but also having the citizen be part of the implementation and delivery of government services. This form of empowerment can ensure citizens take on some level of responsibility in government and governance, ensuring a certain level of accountability to the execution of policy and regulation.

Emergency use authorisation was the key tool used by the US FDA and other health regulators to approve COVID-19 testing solutions and vaccines. As the pandemic grew, at an accelerated-pace health regulators had to move faster than normal to approve solutions that would help combat the pandemic. This threw a major spotlight on the speed regulators can move when faced with major challenges.

The role of regulators is to regulate an industry, whether that is financial, health, or the environment. They act as the police officer of an industry to ensure players within that industry are following the right rules and regulations. Regulators are unelected bureaucrats and are mostly executives from the industry they are regulating, which obviously creates conflicts of interest.

Another key issue with regulators is that they stifle innovation. The best example is of the current issues with the gig economy. Uber and other companies that are technology platforms which connect independent contractors with customers instead of hiring drivers are in a constant fight with transport regulators. The best example is Transport for London in the UK who cancelled Uber's license.

Regulators have failed in most cases, whether it's the global financial crash of 2008 where they failed to regulate the credit defaults swaps market, or environmental regulators who have failed to reduce industrial emissions. Regulators still are fighting the emergence of new cryptocurrencies such as Bitcoin instead of embracing them and accelerating their adoption.

As we build back better, regulators will need to be ecosystem enablers and work with citizens and businesses to develop regulation which protects citizens but also promotes and accelerates innovation and builds the digital economy. Many regulators have created regulatory sandboxes or regulatory labs. These platforms promote experimentation and innovation through private and public partnerships to create new digital economies. The role of a regulator in the new digital world order would be to protect the public good, be accountable, and involve citizens in decision-making through new innovative representative models, but also create new digital economies. Some regulators through government-related entities, especially in the Middle East, have created state-backed companies which drive forward with innovations which create new economies and industries controlled by the state.

In the new post-pandemic society, new regulations will need to be created for artificial intelligence and data management, amongst others. The speed of decision-making and the creation of new regulations will require increased speed for economies to be able to accelerate innovation and create new digital economies and industries which generate jobs. Regulators and regulatory enforcement will be a combination of citizens, representation, artificial intelligence, and businesses. The role of the regulator in the new digital world order is to be a digital economic accelerator one that creates the enabling foundations for new companies' entrepreneurs to enter a market with new innovations. At the same time, regulators will ensure competition by reducing barriers to entry, enforcing fair pricing, and protecting the citizen and their rights. They will ensure that regulation is in lockstep with emerging technologies that benefit society and create jobs. As we build back better, regulators will have a key role but should not be a bottleneck to progress or change. They need to be an enabler of transformative change. The countries which can embrace this model of regulators being accelerators of the digital economy will see unprecedented growth in foreign direct investment within their digital economic sector it will not be just the place, but it will be policy and regulation which will attract investors to invest in the digital economy. This does not mean regulators should have a laissez-faire approach in enforcing regulation. Regulation enforcement must be transparent within the public eye and data driven, increasing the awareness and performance of companies who either are following regulation or breaking it, for anti-competitive behaviour, price controls, etc. Citizens in this through new innovative representative models, can be the regulators and be given the right to vote and grade performance of companies, which directly impacts a companies' bottoms line. This type of performance monitoring model will ensure that companies are truly accountable to the citizens, who are also, in most cases, the end consumers as well. Accelerating the digital economy will need to be the key priority for government and regulators in post-pandemic reconstruction. This will enable the creation of new jobs and new sectors of the economy by empowering new entrepreneurs, especially youth and women, through new regulation, making it easy to register a company, open a bank account, and access open government data. Creating these open data platforms for each industry, e.g., health, finance, agriculture, etc., and then promoting start-ups to create new products and services around this data will generate jobs and new economic value within a regulated environment which embraces new technology through policy and regulation. This will attract national and global companies to set up. The creation of free zones focussed on promoting new digital economic sectors, along with making it easy to set up a company and regulations which promote innovation will be key platforms for governments to build back better in the new normal. As the digital economy becomes the new growth engine for national economies, securing talent will be a key component.

The justice system is based on judges, jurors, and lawyers. This structure of justice has not changed for hundreds of years. The justice system in general is highly inefficient wherever you go. The cost of legal fees can be considerable, and a judgement could take weeks, months, and sometimes even years. As we build back better, there must be a new model of a justice system for the new digital world order, one that is fair, just, affordable, and efficient.

The current system is dependent on convincing a jury or judge that an individual is guilty or not based on evidence but also how good your lawyer is in making your case. This can lead to innocent people being sentenced to prison and even sometimes death for crimes they did not commitment. Legal precedents are always used when making a case, and sometimes these precedents are hundreds of years old. Another key consideration point is most criminal justice systems are based on English common law, especially in countries which were part of the British Empire. Human beings always have a bias; however, AI can only decide based on data. Leveraging AI to support and speed up decision-making will be key in the post-pandemic society. The use of IOT sensor-based technology to assess if a witness or an accused is lying with near 99 percent accuracy will be able to ensure quick and speedy justice. Having an AI brain which can process billions of data points from millions of legal cases and precedents and formulate judgements will remove a layer of the legal system which is inefficient and ineffective. An AI-driven legal system can advance and enhance progress, development and equalise justice for all and not just a few. However, qualified judges will still be required to evaluate the judgements the AI recommends.

Such a system would no longer require a jury to pass judgement and would also reduce the role of lawyers. This cloud-based courtroom would enable judgements to sometimes even be passed within hours, maybe even minutes, depending on the crime. This new form of justice would move society forward and bring back trust in the justice system. A fully transparent justice system which is data rather than money or even emotion driven would increase the rate of verdicts and reduce unfairness in some countries where the justice system is outdated and does not give equal access to the common man who is also the victim or the accused. The fundamental right of equal justice in this case is not practised in real life. The justice system has become something which protects the most powerful in society rather than the most vulnerable. As we build back better, a more equal and just post-pandemic model, the justice system, laws, and regulations which undermine law, order, and society need to be just and focused on creating harmony and happiness between all sectors of society. The cloud-based courtroom will automate and support verdicts. One of the key foundations of the Anglo-Judaic period of enlightenment has been a transparent and efficient judicial system that people trust is fair. When a country has a fair and just system of justice where the rule of law, which is supreme, are the societies which succeed and grow. Creating trust in the judicial system where people and businesses understand there is a nonbiased and fair system of justice is fundamentally what has driven the progress of the Western world. If countries are to progress in the new digital world order, having a strong, free, fair, and efficient justice system economically and socially will be critical.

Chapter 11
A Global Digital Reserve Currency

The Bretton Woods system is a financial system adopted by Western governments after the Second World War. It created the International Monetary Fund and the World Bank and ensured that currencies were pegged to gold and used the US dollar as the global reserve currency for global trade. In 1971, the US ended the gold peg to the US dollar and convertibility to gold. From then on, the US dollar as a fiat currency would essentially be the peg for global trade that would underpin a global financial system.

It continues to this day, where for countries to sell and buy products, they need to purchase them in US dollars. This requires having a reserve of US dollars through selling more to the US or through purchasing US treasury bills. The more demand for US dollars means that the US Federal Reserve can print more money, being the reserve currency of the world and not being pegged to any type of physical asset such as gold. However, since the start of the twenty-first century, the Euro and the Chinese RMB (renminbi) have also emerged as reserve currencies for international trade.

Over the past ten years paper-based money has become digitized. Our wallets have become a mobile app on our phone powered by Google, Samsung, Apple, or We Chat. Transactions happen either through a tap of a phone on a payment terminal or a scan of a QR code. The use of cryptographic and Blockchain technology has also created new digital currencies. Some are pegged to a fiat currency such as Tether. These are known as *digital stable coins*. Some have essentially become digital commodities, such as Bitcoin and Ethereum, with Bitcoin now becoming like a digital form of gold due to its scarcity.

Digitization of asset ownership in the form of shares has also emerged through digital security tokens. Ownership can now be digitized in the form of tokens and held in a digital wallet and traded on digital exchanges.

A whole new digital monetary system has emerged, powered by Blockchain and cryptographic technologies. This new system is outside the control of the traditional central banks and the mainstream financial system. Anyone can now create a digital currency, and it can use trust in the currency and the community, like how fiat currencies are underpinned in the trust in a central bank and government. Digital currencies can also be underpinned by real physical assets, such as commodities like gold and oil, or pegged to a fiat currency. Central banks, governments, and big banks are being challenged by new fintech innovations within both the banking space: better and more nimble digital banks, digital wallets powered by Big Tech, and new digital currencies.

Due to the fact these new challengers don't have legacy infrastructure and see opportunities of disruption to the existing financial system, where even cross-border transfers can be instant without exchange fluctuations, governments are looking to create their own central bank–backed digital currencies. During the pandemic we were encouraged not to use physical cash and instead use digital money through our plastic cards or our digital wallets.

China is now leading the way in using Blockchain and cryptographic technology to develop its own central bank–backed digital currency. This would be existing legal tender but would be fully centralised and traceable using a centralised Blockchain. In the digital world order, we will move to digital money underpinned by a central bank–backed global digital currency, driven by a new digital Bretton Woods. It will be a global digital reserve currency that is backed by a combination of commodities and a basket of fiat currencies. This will give it stability and interoperability. By creating a global digital reserve currency, other digital currencies can be created and use this as a peg. Brands can create their own currency, and so can social media influencers. These currencies can be pegged and convertible into this global digital reserve currency. As this digital reserve currency is based on Blockchain, it will be fully transparent and traceable, meaning it will put an end to money laundering and foreign exchange fluctuations.

This new global digital reserve currency will be linked to an individual's self-sovereign identity and exist within their digital wallets. Central bank–backed national or regional digital currencies will be pegged to this global digital reserve currency. Each country that signs up to use the global digital reserve currency will have to allocate a percentage of their state's assets, such as fiat currency, gold reserves, and other commodities, to ensure stability and trust in this new digital monetary system. This could be in the form of digitized securities such as sovereign bonds which can be traded on digital exchanges.

Part of the creation of a new digital Bretton Woods is the opportunity for small and medium-size business to access global investors and global markets through issuing digital shares. Digital shares will allow small and medium-size business to access liquidity pools globally and give investors an opportunity to invest in private business through digital security offerings. These digital securities are held within a digital wallet and exist on a transparent Blockchain, ensuring clear records of ownership. They can also be traded on digital exchanges like the trading of stocks on a traditional stock exchange. This would allow private businesses to raise capital from the public markets in the form of crowd funding. Automation will replace millions of jobs, and consequently, new business and new industries will need to be created, which will require access to capital. By creating a global market using digital securities, anyone with an idea or an existing business can raise capital through a digital securities offering, where share ownership is digitized and fully transparent using Blockchain technology which is immutable and cannot be disputed. This level of efficiency and transparency allows any type of investor to invest with confidence. As we build back better and rebuild a new, better, and more inclusive global financial system using the latest innovations and technologies, it will accelerate the creation of new jobs and new industries that drive the post-pandemic reconstruction and the new digital world order. The creation of a new global digital reserve currency will require global oversight. The creation of the euro was a big experiment in the creation of a single market with a single currency. This type of monetary union does not exist anywhere else on the planet. There are many things from the European monetary union example which can be used in the development of a single global digital reserve currency. If we look just at cross-border e-commerce transactions, by paying in one global currency, it would mean less cross-border exchange fees and bank charges. Even banks could create their own currencies using the global digital reserve as the primary peg. Such innovations in money have not occurred since the start of central bank-backed sovereign fiat currencies. Such innovations would also create new economic blocks. For example, the Chinese's One Belt One Road initiative would promote the use of the Chinese central bank-backed digital currency. Having a single global digital global reserve currency that all countries can use allows for harmonised standards for trading. New national digital currencies can also be created using the global digital reserve currency as a peg.

This radical change in the global financial system away from the US dollar will create pushback. However, as we emerge from the pandemic to a multipolar world where there is no longer one global superpower, but trade, commerce, travel, etc. continue in our global village, such thinking and discussion are around a more centralised but global digital reserve currency which is single-country agnostic and takes into consideration the growing powerful economies of Asia, the Middle East, and South America.

The two disruptions as we build back better will be the increased use of digital wallets and digital currencies and the creation of a multiple polar world which is no longer dominated by the US dollar. The US dollar, like other major currencies, will still be a major part of the basket of currencies and other assets which the global digital reserve currency will be pegged against instead of being the only one.

Digital Tokens are programmable pieces of software which can hold some type of value but are not digital currencies. These digital tokens are recorded on a Blockchain and can exist on a centralised or decentralised Blockchain, private or public. Digital tokens can take the form of a *smart contract*, which is a digital version of a paper-based contract. This is ideal in supply chains where there are multiple parties handling the same goods from manufacturer to logistics provider to retailer. They give full visibility of where a product is within a supply chain and ensure that each party has fulfilled their role before it is handed over to the other party. Smart contracts are particularly useful where multiple systems and IOT devices are involved, as they provide instructions from one machine to another. Think of a production line, where different machines undertake different tasks. A programmable smart contract allows machines to talk with each other, ensuring that each machine has completed their tasks. Smart contracts can be stored within a digital wallet, making it easier for them to be accessible. Such innovations can ensure any type of paper-based contract can be converted into a smart contract, whether that is a title deed, a sales purchase agreement, etc. As these smart contracts are registered on a Blockchain, they are immutable and therefore cannot be tampered with. This significantly reduces fraud, with clear ownership rights linked to people's ID. As we will build back better, this type of digital token can revolutionize legal contracts, from property ownership to records management, reducing disputes and allowing IOT smart devices to talk to each other and execute contracts effectively.

Non-fungible tokens (NFTs) have emerged to allow people to own a digital or physical product, such as a piece of art or a book. Non-fungible means they cannot be copied. This is a great way to stop piracy of music, videos, and other forms of digital entertainment. In the new digital world order, an individual will own their own data and will only share it with whom they want in exchange for money or rewards. NFTs, which are designed to allow individuals to share data, will revolutionize the advertising industry, allowing individuals to sell their NFTs directly to advertisers through NFT data exchanges. Advertisers can then personalise advertising messaging based on the individual's behaviour by reading the NFT. This type of advertising gives the individual power over their data and allows them to monetize that through a permission-based data tracking system.

NFTs, by revolutionising ownership, will change the open-source economy by allowing inventors to monetize their inventions rather than giving it away for free. Programmable tokens will also revolutionize the rental and lending market, allowing people to own something for a defined period rather than permanently.

The token economy will form the basis of the digital economy, allowing companies and individuals to develop new token economic models of ownership and storing and exchanging value from physical and asset ownership through digitising contracts, all registered on a digital ledger based on Blockchain.

The fractionalisation of real estate assets allowing anyone to own a fraction of a real estate asset through buying asset-backed digital tokens will make real estate investment more accessible and inclusive. New models of owning a house through rent-to-buy schemes and owning real estate investment through digital tokens, where all paperwork is transparent, immutable, and secure using Blockchain technology, will also reduce fraud and increase trust within the real estate market. In nations where population growth is outstripping the supply of housing, the fractionalisation of real estate assets using asset-backed digital tokens can attract investment but also allow the common man to own a piece of the property they rent, thus allowing the masses to benefit financially from a sector in which they could never be able to take part in. Owning something through a digital token which can then be traded on a digital token secondary market will become a major funding source for companies but also a source of income for people. Governments can also tokenize infrastructure projects, allowing their citizens to participate in economic development and therefore citizen fund massive infrastructure projects allowing local and global investors to invest through fractional ownership. This type of funding model for state backed infrastructure projects removes middlemen and brokers, reduces fraud, waste and abuse but also allows the common man to invest with a few hundred dollars. This type of model can also promote community development where the members of the community can self-mobilize and issue digital tokens for an infrastructure project within their community, which improves the quality of life of their community but also, they are able to gain some type of return back to the investors living in the community as a secondary income.

Chapter 12
The Energy Transition

Human progress has always been enabled by how we can harness and use sources of energy. The first three industrial revolutions were based on how steam energy could be harnessed to accelerate industrialisation using hydrocarbons such as coal, gas, and oil, as well as hydroelectric power and nuclear. Massive, centralised electricity grids were created that could transport electricity through copper wires to every home and every business by massive power stations. The energy which was produced polluted the atmosphere and water sources. The production of excess levels of carbon dioxide in the atmosphere and other harmful greenhouse gases driven by energy production and automobiles contributed to global warming and climate change. The use of hydrocarbons led countries to rely on other countries for sources of energy, such as Middle Eastern countries. The transportation of hydrocarbon energy through a massive network of pipelines led to energy politics, which created wars and massive amounts of instability across the world, most notably in the Middle East, which would become the hub of hydrocarbon energy. Energy dependence and securing hydrocarbon energy became the global power play of the twentieth century, the US dollar become the currency for the trade of hydrocarbon with the emergence of the petrodollar, and oil become the main source of energy for industrialisation and the advancement of society. As everything was linked to oil, OPEC (Organization of the Petroleum Exporting Countries) became a very powerful cartel. Western nations became consumers of OPEC members' oil, and so a love-hate relationship started between the advanced Western nations and the mostly developing OPEC nations began nearly fifty years ago. Over the next fifty years this governed politics, foreign policy, and security. To gain control over supply, consumer nations tried to ensure they had influence and power over producer nations through economic, political, and military power. This led to producer nations falling short of developing and diversifying their economies due to their heavy reliance on hydrocarbons as the main source of revenue.

In 2016, the world signed the Paris Climate Agreement. The goal was to move to a net zero carbon world, giving each country a target to transition away from hydrocarbon energy to sustainable green energy such as wind, solar, and hydroelectric. Electric cars, with the disruption of Tesla cars, signified a major transition from transportation being powered by hydrocarbon to electric. As we build back better a cleaner more sustainable energy production model, we will see major changes to the relationship between current producer and consumer energy nations. We may also see the emergence of new energy producers. For example, countries which can generate all their energy from solar power can potentially export that power to other nations. The same goes for wind, and the same goes for hydroelectric. We also see the emergence of a smart grid where a house or building produces enough energy to sustain itself and gives any excess energy back to a smart energy grid. This means any house or building becomes a power station through sustainable green energy which is powered by clean batteries and IOT sensors. Countries like the UAE and Saudi who were and still are the main producers of hydrocarbon energy are now building some of the world's largest solar parks to harness the power of the sun. As part of the energy transition will be the increased use of recycling materials and energy, the circular economy will focus on reusing non-biodegradable materials in ways which create new recycling economics. This will be even more important for electronics. Electronic waste will be the single biggest issue to tackle with the growth of digital technologies and the constant upgrading of devices. Digital waste will be major problem, and new models of sustainable development and new materials will be required to create sustainable electronics.

The importance of sustainability net zero carbon and the circular economy will be key themes on which investments will be made. Investors through the environmental, social, and governance (ESG) model will evaluate companies through this lens. If companies don't adhere to or follow this model, it will have a fundamental impact on gaining investment.

Multinational companies will need to change the materials they use for their products, whether it's the plastic bottle of your shampoo or how our smartphones are made. This will not just impact the attractiveness and value from an investment perspective, but consumers will also evaluate brands, and their purchasing decisions will be focused on how a company behaves, and if it follows environmentally sustainable practices. As we build back better, the bottom line must be green, sustainable, and responsible. A green tax will need to be imposed for companies which do not adhere to new green measures, from how they manufacturer products to how they use energy and how they invest in growth and contribute to the communities they serve.

A new race will emerge, which is the race to secure rare earth minerals such as lithium, which is used for battery production. Countries which have access to raw earth minerals will become the new producer nations, the new oil economies. Oil will be replaced by lithium as energy changes from hydrocarbon to battery-generated electric energy. Currently China is the world-leading producer of raw earth minerals. However, as the world transitions to electric energy, the discovery and mining of raw earth minerals and securing these sources of new energy will lead to changes in energy politics and policies and as a consequence new nations will become global energy players.

Chapter 13
The Digital Economy

People can set up an online store in minutes and sell anything from anywhere. A global supply chain of logistics, warehouses, e-store templates, drop-ship products and services, marketing channels, and freelance communities exist, which anyone with an internet connection can use to sell their products and services at scale globally. This means anyone can compete with a global multinational. In fact, not just compete, but also be faster to market, be more agile, and drive scale without a massive corporate structure and inefficient decision-making processes. Multinational corporations must now compete with multinational sole traders who have a drop shop selling whatever, access to freelance designers, and marketing and social media platforms that give them access to billions of people.

With the COVID lockdowns, companies realised that it wasn't necessary for them to have big physical offices, which meant employees were connected via the cloud, video conferencing with their colleagues. But what is the need for such a large corporate structure if as a multinational sole trader, you can sell the same products and reach the same customers using different supply chain logistics and freelance platforms.

With the advent of the social media influencer who has hundreds of thousands if not millions of followers, it became much easier to market products and services through endorsements from these social media stars.

This meant you didn't need millions or even billions of advertising budgets to reach your target audience quickly and effectively.

What the COVID lockdown also brought to the forefront was e-commerce, which was already accelerating, with malls and traditional retailers in decline as more and more people shifted to online buying.

This also meant that not just retailers, but also manufacturers, had to shift to a digital business model. Some companies had already started to shift, but some started to because of the need to stay in business.

However, these large retailers or large manufactures have lost a major advantage due to e-commerce. Anyone can compete with them, so the multinational sole trader who can set up a drop shop, white label a product, use social media influencers, market a product with zero inventory, and ship from source can compete, and even outperform, the biggest global multinational.

The scale, uniformity, growth, and access to new markets, distribution channels, and logistics, which were the hallmarks of what made the global multinational, are now no longer a competitive advantage. In the new digital world order, our goal should be to create a billion new multinational sole traders globally. This new empowered entrepreneur is creating the new McDonald's, the new Nike, the new P&G, the new L'Oréal. In fact, if we look back in history, the majority of global multinationals started as sole traders and then expanded, driving mass markets, mass production, and mass-scale economics.

We now go back and forward, with the opportunity to create a billion new digital cottage industries, which drive the new digital world order. Here, the shareholders are the customers or even the community. They are focused on prosperity and not just pure profit taking, giving back, and uplifting the community and the wider environment. They are driven by the purpose of delivering a return on happiness, which translates into loyalty and sales rather than shareholder value, delivering self-holder value and community-holder value. These will be the key measures of success of the multinational sole trader, as they become the new engines for growth that drive the post-pandemic global society.

Today, farmers in Africa could deliver fresh coffee through Instagram, anywhere, within twenty-four hours, due to drone delivery and logistics. They can cut out the middlemen and create a direct link between themselves and their customers, without the need for even going to retail. This is not a pipe dream. This is a reality. Everything needed to do this exists today, and as such, this type of new model of economics which brings producers closer to consumers will play a major part in uplifting billions of people globally out of poverty.

The multi-national sole trader will be the engine for economic recovery and the creation of new digital economies. As part of post-pandemic reconstruction, governments should focus on how they can allow these new digital-driven sole traders to quickly set up their business with a bank account. Currently, the creation of a new business, getting a license and then getting a bank account, can be a bureaucratic and inefficient and expensive process which discourages entrepreneurs from setting up a business. In the new digital world order, it will be not the place but the policy that will attract businesses to setup. In Estonia, the ease of which anyone can set up an e-business license online is a great model for other countries that want to create and attract foreign direct investment or create new local entrepreneurs and businesses and build a digital economy.

The digital economy is about digitizing existing industries and creating new industries, developing skills for the jobs of the future. The digital economy is about leveraging data and technology to develop new products and services.

Governments will need to create open data platforms which allow businesses and entrepreneurs access to data, whether that is social, credit data, identity information, health information, etc. This allows for innovation and creativity, thus allowing the creation of new digital economies. As the digital economy becomes the new growth engine for national economies, securing talent will be a key component. Tech talent will be a key national resource in post-pandemic reconstruction and the development of a national digital economy. However, not all countries will have that national tech talent. Countries that have always relied on foreign workers, such as the Gulf Arab countries, will need to not just promote a national agenda to develop local tech talent but also secure tech talent from producer countries if they are to accelerate their national digital economic agenda. Similar to the strategy of food security, where Gulf Arab countries have successfully executed by securing agricultural land in producer countries for their own national security, a similar strategy will need to be applied by any country with tech talent producer nations. Highly qualified and cost-effective tech talent from countries insuch as Eastern Europe the Indian Subcontinent and developing Asia will become the new producer nations for the digital economy.

Each country will need to create strategic tech talent pools and will need to have a strategy on how to secure and attract this talent. In the new digital world order as we build back better, tech talent security will be a key area of focus for countries during post-pandemic reconstruction.

The G20 needs a global goal of creating a billion new businesses over the next ten years enabled by data, automated payment systems, and global supply chains which anyone can connect into. They need to make it easy to set up a business from anywhere and at any time in minutes with a bank account. This can encourage automation and lessen the impact of the massive number of job losses the pandemic has created. They also must encourage the growth of fast mobile broadband connectivity across the world. For example, Pakistan, which is a country of around 250 million people, has around 65 million mobile users with at least 3G internet connectivity speeds. That is more than the population of the United Kingdom. This digitized user base is an active market for the creation of new digital economies. This is the case in a lot of developing countries across the world. The high-speed internet through smartphones has enabled access and created a platform to create new businesses. Unfortunately, government bureaucracy in terms of ease of doing business along with enabling new companies to gain access to capital has not matched the opportunity in developing nations. The growth of vertical and horizontally connected global platform marketplaces such as Shopify for retail or Fiverr for free lancing, amongst others, has allowed anyone to start a business.

Digital economies built around digital platforms that support the growth of each industry, which act like demand and data aggregators, will allow entrepreneurs to quickly tap into demand with a superior or low-price product. Regulation and legislation to support the development of a digital economy is a critical element of post-pandemic reconstruction. Without this, many economies and countries will fail. Social instability will arise, as people who have lost their jobs with no clear route out will create a burden on state welfare systems. In turn, the lack of economic production will impact government revenue.

The pandemic accelerated the 4th industrial revolution. Some industries will need reindustrialise digitally to be relevant or incumbents will be replaced by digitally focussed companies. For example, auto manufacturers are moving from combustion engines to electric and self-driving, and totally new industries are emerging, such as the data industry, data mining, data analytics, data insights, and data monitoring. All of this will create opportunities for new market entrants and new business models.

The opportunities that 3D printing provides are immense. Produce your own toys, clothes, food, furniture, and even medicines. This new maker movement empowered by a 3D printer and an online store will provide new opportunities for the multinational sole trader to sell and market new unique products which are made to order. Every home has the potential to be a mini factory, churning out products which are personalised and unique to the individual. This moves us from mass production to mass niche production where people publish the designs of their products online, and people adopt, adapt, and produce them at home using their 3D printers, then sell them on online stores marketed via social media channels. Such a new model of production lowers the barriers to entry for new entrants into the manufacturing industry. It also threatens the large global supply chain of multinationals who must now compete with someone who can produce a product and market and sell it just as well and if not faster than a multinational with the benefit being that of personalisation at scale, which is something multinationals with their one-size-fits-all supply chains cannot deal with.

3D printing will also revolutionise medicine, where you can a have 3D-printed pill that is personalised to the needs of your body. From designer vitamins to designer medications, all you would need is a cartridge and a program to run on your 3D printer with the correct programmable formulation. For larger 3D-printable items which cannot be printed at home, a new type of community-based 3D printing service will start, where a community has a large enough 3D printer to produce furniture, food, and other large items based on the specs and the designs of the customer.

3D-printed clothes will ensure that when you buy something, it is tailored to your exact size. This will disrupt the current clothing supply chains, which use low-cost sweatshops in developing countries such as Bangladesh, using low-cost child labour.

The design-your-own, build-your-industry powered by innovators and 3D printing will significantly disrupt the global supply chain and the current models of production.

We are already seeing low-cost 3D-printed homes which can be built in days, and parts of the human body being 3D printed, a new finger or a new lung.

Digital natives are more inclined to develop their own brands and personalities, being distinctive with their own styles rather than conforming to a certain look or brand. The emergence of 3D-printed clothing, where anyone can create their own designed clothes and even a personalised branded collection will drive disruption within the fast fashion retail industry. It will also promote and create new designers, the next Giorgio Armani or Tommy Hilfiger, who are able to use social media influencers to promote their designs.

The disruption of 3D printing will bring to manufacturing will be immense. The speed and efficiency of production and the reduction of capital investment costs along with labour will dramatically reduce the cost of many products. It will also create opportunities for new products which may not have been economically viable in the past.

For governments, the creation of policy and regulation that creates a new 3D printing industry will be an important aspect of accelerating the digital economy.

Investment in 3D printing technology as well incentivising entrepreneurs to create 3D printing businesses. The housing shortages in most countries and the rapid shift of workers from rural to urban areas, especially in large developing nations, has created uneven development. In the new digital world order, everyone is entitled to a roof over their head, and low-cost 3D-printed houses within smart connected communities could be the answer to the problem of the housing crisis.

Mass personalisation at scale, which 3D printing will create, will need to use sustainable and recycled materials. This combination can solve multiple problems. Having 3D-printed homes made of recyclable waste materials developed at scale is a sweet spot. It solves the housing crisis as well as accelerates the recycling industry.

Having a 3D printer in every home and every classroom will unleash the creativity of the human mind. It will provide children the opportunity not just to learn from books but also to learn from building and doing. As we build back better, this can be a major building block to accelerate the creation of new industries. Due to the pandemic, millions of people will have lost their jobs. Promoting and subsidising the use of 3D printers can spur new economic activity and a new digital industry, from a new type of construction industry to new medical and body parts made of 3D-printed materials to 3D-printed foods creating new jobs, promoting new skills.

Chapter 14 – Economics of Happiness

The gross domestic product of a nation is a measure of economic productivity. However, economic productivity does not give a true reflection of how well a country is doing and how happy its citizens are. Shareholder value in the form of a company's stock price is not a true reflection of how well a company is performing and if its customers and employees are happy. The UAE is a great example of taking citizen happiness as a strategic priority by dedicating a Ministry of Happiness. The breakdown in Western society because of COVID-19 and the lack of trust between citizens and their elected officials is fundamentally down to trust and happiness. In the new digital world order, countries should focus more on gross domestic happiness (GDH) instead of gross domestic product (GDP) as the key measure of prosperity. Happy citizens are productive citizens, which leads to economic and social benefits.

Companies, instead of focusing on their stock price and delivering shareholder value, need to focus on delivering happiness value for their customers and employees. With this focus, employees will become more productive, and customers will become more loyal, therefore increasing economic prosperity. In the post-pandemic global society, it will be the economics of happiness which will drive businesses and economies forward, increasing prosperity and happiness. Happiness can be measured by sentiment analysis. On our social media feeds, we no longer just "like" things. We can express love, sadness, and multiple other emotions. It's the same for when we text or WhatsApp using emojis. Emojis have become a new universal language and expression of our emotions digitally. We use emojis every day, and they have become a unifying digital language of the twenty-first century. Happiness has many facets: economic, social, cultural, and emotional. These drive human behaviour and shapes our thoughts, needs, and wants. Money doesn't buy you happiness, but in our capitalist view of society and culture, money, and lots of it, always seems to equal happiness. In many cultures, love, family, health, and community are drivers of happiness. In the new digital world order, a customer experience or customer journey can be fully quantified with data points that can measure sentiment and emotion. Brands using digital technologies are seeing a fundamental shift in how consumers experience their product and service offerings. Due to the digitization of the customer journey, where any product or service is on an app or a website, having a physical location or physical in-store distribution is no longer a competitive advantage. Your competition is a click or a search away in the new digital world. Each product or service can be rated and reviewed. On each article and post, instead of just giving a "like," you can communicate other emotions using emojis.

Each step of a customer's journey can be mapped. It's the use of this vast treasure trove of data where companies and governments can gain sentiment and emotional intelligence and analytics. Therefore, consumer happiness can be gauged and segmented at a very individual level.

As we build back better, happiness, both as a citizen of a state and a customer of a product or service, will be a key measure to define success. Loyalty is a form of trust and happiness with a product and service. It takes a long time to build and a second to break. Trust and happiness are two sides of the same coin. Happy customers are loyal customers, and happy citizens can create a positive and social environment. If we look at the gross domestic product measure of a country or the profit and loss of an organisation, there is no measure of happiness in any of these measures. Price and quality do not capture other key factors, such as social responsibility and environmental friendliness or how an organisation is contributing to the health and well-being of communities. A company can be making lots of money and have a high stock price but not have happy or loyal customers or employees. This can become a medium- to long-term threat, as technology changes and disruption can come in, disrupt an industry, and make the incumbents redundant because they never had happy or loyal customers or employees. The same can be said for governments where a lack of trust and happiness can lead to an election defeat or protests in the street and a toppling of a government.

A global happiness index should be created for both countries and businesses which can then be used as a benchmark to measure success. For both governments and businesses alike, it should be a combination of internal employees' happiness and external customer/citizen happiness. Such an index can benchmark companies and governments, which can show and drive a cultural shift from a stock price to a happiness index, from gross domestic product to gross domestic happiness.

With the vast data lake of sentiment and emotional data available from social media, online polls, and sentiment analytics, both companies and governments can use this data to develop a happiness index. They can also use it to reward along with penalise companies and even government ministries who do not conform to benchmarks, naming and shaming them in public. This type of happiness index along with putting a public spotlight on good- and bad-rated entities will drive behaviour changes amongst leadership teams to make happiness a strategic KPI. Currently happiness is not a KPI that is captured, and boardroom leadership teams are not measured against happiness. Happy customers are loyal, and this turns into financial reward along with happy employees, which means loyal and productive employees. Productive employees also contribute to financial reward.

As we build back better and gain the ability to measure happiness through multiple different digital channels, happiness as a key measure of prosperity and well-being can be captured and measured on.

Gross domestic happiness in the new digital world order will be the key measure to measure the success of a company or a government. Such measures should contribute to social stability, happy employees, customers, and citizens, creating an environment of community stability and cohesion. Mental health during the pandemic due to lockdowns became a major issue. Post-pandemic this will be the single biggest health issue after COVID. Employees and governments will need to double down on developing and promoting solutions and programs that support mental health and wellness. Mental health is still often seen as taboo in all societies. It something that cannot be seen but can be very dangerous. Often people don't want to admit they are depressed or suffering from mental health issues. Governments should mandate mental health and wellness programs within the workplace and schools. The use of technology-driven solutions should be incentivised to allow start-ups to develop solutions such as AI-driven digital humans that can help develop mental health and wellness solutions which can track mood and provide advice and solutions. As we build back better, the mental health of populations due to continuous disruption of normal life through lockdowns has taken a major toll on individual mental health, straining relationships. This could be a silent and invisible pandemic after COVID-19, which can lead to dangerous consequences for society if it is not prioritised and acted upon properly by both employers and governments. It can reduce the output and cause instability within the workplace. For governments, it can lead to instability, protest, and violence, which could lead to unexpected and difficult consequences for the state and for businesses.

Chapter 15
Universal Health Income

The pandemic has shown that human health is interlinked across borders. If one of us gets sick, that can have an impact on the entire world's population. The velocity of the spread of COVID-19 showed how now and, in the future, we will be dealing with a new type of biological enemy, one that has no religion or colour and is invisible. We live in the new normal, a new normal where we will be living with pandemics and epidemics for a long time.

COVID 19 brought to the forefront the importance of health. Access to healthcare and health security will become the dominant priorities in a post-pandemic society. The importance of having a robust healthcare infrastructure including biodata, bio-intelligence, and bio-surveillance will become fundamental components in ongoing future pandemics and epidemics.

The only way to fight an invisible enemy is through biodata and bio-intelligence. This will require the development of a massive and global biodata solutions, like the Facebook for health data where billions and billions of people's health data is stored. However, people will need to give permission for anyone to access this data. As health data becomes extremely valuable in a post-pandemic society. However, people will need to be motivated to share such personal data.

A universal healthcare income can be created, which is provided through a mixture of government and non-government providers, like a universal income but linked to your health. In the new digital world order, this digital currency, based on the principles of Blockchain, will be a stable coin that will be recognised as a universal currency, which can be used in exchange for goods and services just like a normal fiat currency. It will be recognised by central banks as a global legal tender and be linked to the global digital reserve currency to give it stability and be accepted universally. Universal health income will be given and will act as a social welfare benefit. For motivation, sharing health data or making sure you have taken the right vaccines will give an individual bonus income, providing motivation for people to be healthier and ensure they are vaccinated. This type of income will also provide benefits, such as an online marketplace which gives people access to universal healthcare, telehealth providers, discounts on purchasing healthy organic food, and even financial products such as insurance and short-term loans and credit.

The universal health income will act as a universal basic income for everyone over the age of eighteen. It will provide the benefit of a global safety net for people but also provide biodata and bio-surveillance intelligence to keep populations safe from current and future pandemics and epidemics.

We may also see ongoing voluntary health monitoring using biochips which are inserted under the skin. In exchange, people volunteering for them will receive additional health income.

By creating a global social security net, we can focus on providing universal health and a basic income to anyone over the age of eighteen globally. This would be a combination of private and public sector partnerships regulated and governed by the people, which is uniform across the globe with some localisation. This is not impossible. If governments can learn from the tech industry, who have billions of users generating billions of data points, it's possible.

In the new digital world order having a global universal digital currency which acts as a social welfare net driven by health will be one of the key pillars of the post-pandemic global society. A universal health income which acts as a universal basic income, will be a global safety net available to anyone with a smartphone and a digital identity. In this system, a universal health income wallet will contain your universal health coins. This digital currency will be given monthly no matter if you have a job or not. It will be used to spend on health and wellness-related activities. If we look at

COVID 19, vaccinations will potentially now need to be given every year or even twice a year to gain some type of herd immunity and lower the death rate and hospitalizations. How will governments pay for this? Will they continue to buy vaccinations and give them for free? How long will people need to pay for PCR (polymerase chain reaction) tests. The current model is unsustainable, and therefore a universal health income needs to be funded by G20 governments and private-sector multinationals, especially insurance companies and foundations. The creation of a universal health currency will provide opportunities for innovations within the health insurance space with a fixed monthly universal health income. A part of this income can automatically be allocated to a global health insurance program focussed on dealing with current pandemics, such as COVID-19, and potential future ones. The pandemic has further exposed the inequality in vaccinations, testing, and the availability and infrastructure to support it. The combination of lockdowns and their impact on economic activity and increased spending on healthcare has increased the level of debt for governments in both developing and developed countries. The way a virus can spread and mutate across borders means a basic universal health income which includes a global health insurance program will not only provide a safety net for people, giving them a basic form of income to survive, but it will also stimulate the economy along with creating an innovative approach to ensure health coverage for everyone. This would help further manage pandemics such as COVID-19 more effectively.

Since the start of the COVID-19 pandemic, there have been calls for a universal basic income due to the impact it has had on the economy, creating mass unemployment due to the reduced economic output. To stimulate the economy, programs have been implemented, such as a furlough program where the UK government subsidises the wages of private-sector employees or in the US where the government is giving everyone a welfare cheque. Such solutions only create a short-term economic stimulus and do not solve the problem and burden which governments will face in post-pandemic reconstruction. As we will build back better reconstructing the social welfare net which was created after the Second World War in the West and has been reapplied in many other countries has become a burden on state finances. Redefining this social contract is a fundamental component in the new digital world order.

We will never really know how and where COVID-19 came from. Did it come from a bat in a market in Wuhan? Did it come from the Wuhan Bio-Lab?

Is it man-made through gain of function, or is it an evolving disease that got passed from a bat to a human?

However what COVID-19 has shown is the power of a respiratory disease to create destruction, chaos, and death at a global level. All the guns, missiles, fighter jets, and nuclear weapons have been ineffective and redundant.

The only way you can win a war with an invisible enemy is through biodata and bio-intelligence.

Unfortunately, building bio-weapons doesn't require many investments, and they essentially can be deployed by human carriers. When you can turn a human being into a bioweapon and create an army of carriers which can spread a disease across populations and borders quickly, we are now moving to the next generation of weaponry and warfare. This can be conducted by state and non-state actors. After the second world and the nuclear bombing of Hiroshima and Nagasaki, we entered the nuclear weapons race with the advent of the cold war. Nuclear weapons guaranteed mutually assured destruction and hence have not been used in warfare, due to the very fact they are the weapons of last resort. If we view COVID-19 as a bioweapon, it has created so much destruction and chaos that it has done as much, if not more, damage as a nuclear weapon.

The biggest threat to the new digital world order and a post-pandemic global society will be bioweapons.

They will be the main weapons of non-state and state actors.

To manage and control this would require the combination of treaties like the Nuclear Non-Proliferation Treaty but for bioweapons, the active monitoring of key substances and their supply chain, and the creation of global bodies like the IAEA (International Atomic Energy Agency) but for bioweapons.

The universal health income, the quantified self, will give active surveillance using biodata and bio-intelligence within a "test, track, trace, and isolate" global framework.

COVID-19 showed what chaos the use of a bioweapon can create. Just like after World War 2, when the biggest danger to humanity was nuclear weapons, in the post-pandemic world, the use of bioweapons by nonstate actors will represent a clear and present danger to our global society.

This must be taken seriously, but also there needs to be global collaboration and sharing of data in real time to ensure that we stay one step ahead. Defence has changed from tanks and missiles to biodata, bio-surveillance, medications, and vaccinations. Access to universal healthcare and health security will be the two key fundamentals of the new digital world order and building a successful and safe post-pandemic global society will require the combination of technology and collaboration between countries, institutions, and the private sector.

New capabilities and new industries have started to develop because of the COVID-19 pandemic. Prevention, detection, and diagnosis together with "track, trace, and isolate" are the key pillars to developing an effective defence against current and future pandemics. The use of biodata and bio-intelligence has significantly accelerated the digitization of health data. Most governments have developed their own track-and-trace applications, vaccination passports, etc.

However, a more global solution to this problem has not been undertaken. Pandemics are not localised; they are global and are able to spread very fast. We have seen this with the different strains of COVID-19. Alpha, Beta, Delta and Omicron etc. have originated from different countries but have quickly spread. Governments are always trying to play catch-up with lockdowns.

Bio-surveillance in real time will be the key to preventing, detecting, and containing any bio-attacks in the future.

This will require the use of IOT devices and sensors. Some may have to be put on a person, especially when they travel.

People who travel may require an embedded chip to assess diseases more effectively and proactively stop spread. Such scenarios and solutions seem like a Hollywood movie, but unfortunately it is the time we live in. To get to some type of normalcy, people will need to sacrifice some of their personal space, especially biodata, for this cause but in return will receive a universal health income. Just like we are seeing a lot of countries requiring individuals to be vaccinated for travel or to enter a public space, this should not be seen as an infringement of civil liberties but a consequence of the new normal we live in. This doesn't mean bio-surveillance should go unchecked or allow governments or private institutions to discriminate against people. It will, however, require regulations and rules governing the use of biodata for bio-surveillance in line with national and international security threats.

Chapter 16
Self Sustaining Communities

The human food supply chain is now made up of chemicals, pesticides, steroids, sugar, and fructose corn syrup. This is due to large multinationals who took over how we farm, making it into a manufacturing business to drive scale and speed. This has had a negative impact on our health over the last thirty years. However, there is a massive movement to eat fresh, organic, and healthy. This movement will continue to grow. In the new digital world order, we will see more local community-based sustainable farming. As we reimagine towns and cities, sustainable vertical farms which are self-sustaining will emerge and replace the massive retail malls which have now closed, as people are now buying online. These malls will become sustainable farms powered by clean energy and hydroponics. These vertical farm malls will sustain and produce for their community. Hydroponic vertical farms, which you can have in your home, will create a decentralised farming model where connected hydroponic vertical farms will allow people to grow their own vegetables in their homes. We will also be able to buy produce online directly from our local farms, cutting out big retailers and brokers, therefore connecting the farm directly to the end customer. Using drone technology, we will not just be able to have farm-fresh produce from our local farm, but also we will be able to have access to farms internationally, where a farmer in Africa can receive an order from someone living in London for fresh bananas, which is delivered via drone within twenty-four hours. This radical change in our food supply chain will create sustainability, remove all the pesticides and chemicals, and even create the ability to recycle food waste for farming. Bypassing the traditional supermarket will reduce costs but provide the farmers with more, therefore uplifting communities. Having our own vertical farms in our homes will help reduce the food bills for families. It will create a healthier society, reducing the disease burden because you are what you eat. Cloud-based kitchens that are connected to vertical farms will revolutionize the restaurant industry. Cloud kitchens, vertical farms, and delivery apps combined will mean anyone with a great idea for a restaurant can essentially connect into this food infrastructure and open a business, promoting it directly on social media. The entire food supply chain for cloud kitchens will be localised within the vertical farm malls. Everything will be in one place, using green energy and sustainable and organic farming practices where fresh produce is used to power these cloud kitchens.

The Eden Project in the United Kingdom, which uses massive dome-like greenhouses to grow vegetation, is a model which can be used to turn any location into a place which can grow any type of vegetable or fruit. The use of hydroponics leveraging IOT technology is not just reducing the amount of water used to farm and grow but also the ability to recycle water. As malls close, they can be converted into hydroponic-based, smart, IOT-connected farms which are not just producing free fruits and vegetables but also meatless meat. The meatless meat movement has already started, and the technology to further mass-produce healthy meatless meat in these vertical farm malls can be the main source of food for communities. Where people used to go to a mall to buy mostly processed food, they can now go to the same mall to buy locally produced fresh fruit and vegetables and meatless meat at a low cost. Such concepts will create new types of jobs for communities that have suffered from mall closures.

Multi-use farm malls which also operate as cloud kitchens will ensure that fresh produce is used to provide cloud kitchen owners low-cost, high-quality local produce, which is free of any chemicals or harmful pesticides. As we build back better, we must also eat back better and re-jig our food supply chain. The community-based farming model existed before we urbanised however as we developed food production became a mechanical mass-production industry. As we will build back better, we have the opportunity through this Great Reset to go back to that of smart community-based farming model. A model which will grow produce based on the needs of the community it serves, powered by IOT smart sensors, AI, data and analytics, powered by green energy and recycling our waste back into production.

Such sustainable models exist and are proven to work. The opportunity to convert empty mall spaces into vertical farm malls and cloud kitchens represents a shift in the food production, distribution, and delivery business. Local governments and even mall owners can incentivize agritech companies to provide a platform for independent smart farmers along with food delivery companies to create a new farm-to-table fresh food model. This will connect cloud kitchens, food entrepreneurs, smart farmers, and the end customer within a seamless customer experience, reducing costs and creating jobs for local communities, as well as reinventing and digitising the local community-based farming model.

Post-pandemic reconstruction will require new models of living. Considering pandemics and epidemics will become something we will need to deal with on an ongoing basis, how we design communities will need to be reimagined. The urban model of development adopted after the Second World War, focused on the car, should no longer be the model for the development of new urban models in the post-pandemic world.

Instead of city-wide urban development, we need to focus on smaller community development: self-sustaining communities that are focused on green open spaces, areas for social interaction, and vertical farming where everything is in walking distance.

We will also see in these new models of smart community development flying modes of transportation, thus reducing the amount of space dedicated to roads and cars.

The notion of urbanisation should no longer be about cement, brick, steel, and bitumen (asphalt). These should no longer be the key success criteria. Urbanisation should be smart, self-sustaining communities and carbon-neutral design for mental and physical wellness, wellbeing, and happiness. Technology in these smart communities would be ubiquitous, and access would be through facial recognition and fingerprints, thus even reducing the reliance on mobile phones.

The reason for smart communities and not cities is because the administration of cities or even states has become unsustainable. Breaking up a city or state into smaller communities which are self-sustaining and self-contained and connected increases administration efficiencies.

3D-printed smart homes that are environmentally friendly and self-sustaining using food waste and green energy. Even wastewater is used to feed the vertical farms, which come as part of the 3D-printed home. The 3D-printed home will also be built from sustainable and environmentally friendly materials and will be able to be constructed in weeks.

Community-based collaboration studios will bring together remote workers as well as students who are doing e-learning into community spaces where they can interact socially.

The self-sustaining smart community incorporates all the concepts outlined in the new digital world order and becomes the key physical and digital platform of the Metaverse on which the post-pandemic global society model will be built on. With the ability to create safe, self-sustaining communities which are underpinned with IOT sensors and other smart technologies that are all powered by data and artificial intelligence.

How we design our living spaces and how this becomes key to the environment in which we operate must take a critical role as we build back better. In Western countries, social housing has become a breeding ground for drugs, crime, and even radicalisation; ghettos have been created. This environment has also resulted in a class of people with no jobs or prospects.

The creation of self-sustaining smart social communities with their capability for vertical farming and 3D-printed studios will enable opportunities for job creation through new businesses. The failure of social housing developed post-Second World War may have created massive concrete towers of social housing or mini cities. However, it never focussed on creating or building opportunities for job creation. As we build back better, the model of self-sustaining communities not only provides safe, sustainable, and happy living environments, but it also creates platforms for the development of new industries. New smart communities will need to be developed to jump-start national economies at a scale and speed never seen before to ensure the creation of new jobs through new industries such as 3D printing or vertical farming.

Governments will have to spur economic activity through targeting new digital economic sectors which are radically disrupting existing industries, such as the construction industry, to use these new self-sustaining communities as models of how we can build back better.

Smart cities across the world are being developed. However, the design of any new community now will need to take into consideration pandemics and epidemic protocols. This will require the need for new types of open spaces which are designed to create social interaction but also protection from spreading viruses.

The use of collaboration studios which are fitted out with high-definition holographic conferencing technology within each smart community will allow employees to work and collaborate from anywhere, connecting not just employees but also students, essentially creating a mixed-reality collaboration environment or Metaverse.

3D-created self-sustaining buildings will be made of recycled waste and powered by clean energy. Vertical hydroponic farms will be connected to cloud kitchens, all powered by IOT sensors, biometrics, AI, and 3D printing and holographic collaboration studios. Drone and flying car ports and electric smart transportation pods will be the model of a self-sustaining smart city which is zero carbon with fresh air and green space, creating happiness and wellness for the people that live there. In a post-pandemic society, this will be the model community that will drive post-pandemic reconstruction in the new digital world order.

Chapter 17
Connected Spirituality

Religion will continue to be a key component in the new digital world order, just like culture. Culture and religion are two key elements that make a society and an individual's belief system. Our belief system is influenced by our environment and the information we consume. With the advent of social media, people can spread their own opinions in the name of religion very easily, whether that is good or bad. Religion is an interpretation of written and known scriptures, which can be manipulated to suit a narrative. These narratives are designed to suit a certain goal. We saw this in the first two decades of the twenty-first century where a peaceful religion called Islam was taken over by radicals, and the narrative and message of this religion was hijacked by fundamentalists. This can also be seen in all major religions, Judaism, Christianity, and Hinduism. The preachers of these religions mixed politics of hate with religion. Religion is very personal, it's very spiritual, and it's our connection with God a higher power, a higher belief that is beyond the control of man. Religion is also a big money-making business for these preachers who make hundreds of billions of dollars a year from people who believe in giving, in charity to do well and be good.

In the new digital world order, religious preachers will need to be trained and understand all religions. Due to the power of these digital preachers who can impact the lives of billions of people through their social media channels, YouTube, and podcasts, it is important in the new model of a post-pandemic global society that these people be properly trained and understand all religions. If we look at the Abrahamic religions, Islam, Christianity, and Judaism, there are more things in common amongst these religions than not. However, the preachers of these religions do not have a strong understanding of each other. To rebuild in the post-pandemic era, it will be important that religion, spirituality, and the digital channels of the people who preach are managed effectively, so that humanity can move beyond what divides us and focus on what unites us. For far too long everything that divides humanity can be traced back to religion. However, all religions teach exactly the opposite. They teach love, compassion for one another, so it is strange that throughout human history, religion has always been a point of friction and war.

We are now battling a pandemic war with an invisible enemy, one that does not care what your religion is. If you have lungs, it can get you. So now we must focus on humanity and our survival to combat the current and future pandemics and develop a new model of a global society that can unite and face new invisible enemies.

The modern-day religious preacher needs just a social media channel and a narrative to gain followers. Each preacher expresses their own interpretation of the Holy Scriptures based on their beliefs and narrative. Just like any other social media content creator, the more followers they have, the more money they make, the more they can cross-sell and up-sell and build their brand. However, having the most followers doesn't necessarily mean that their interpretations of ancient holy scriptures are correct. Religion is an interpretation, and it is important that these digital preachers are regulated and qualified to preach. During the war on terrorism, we heard of hate preachers, people who took the literal meaning of what was written down in holy books and without interpretation and contextualisation and its meaning in modern day society were able to justify and radicalise mostly young people to become suicide bombers, all in the name of God. There are obvious points of friction in the world today. Some issues are nearly a hundred years old, if not more, and often revolve around religion. As we build back better, these digital preachers or social media religious influencers need to follow a code of conduct. Social media is a powerful tool to unite the world but also divide it. It is the latter that must have a check and balance, especially when it comes to people which teach and preach religion. These accounts need to be verified, where their qualifications are reviewed for them to preach. This is not an infringement of freedom of speech. It is creating trust, responsibility, and a check and balance to such an important part of society. One of the reasons why radicalisation has occurred is because of foreign policies of Western governments, especially in the Middle East. It is also because of the overbearing nature and voice of religious hate preachers who have used social media to spread messages of hate and intolerance on political and social issues. However, this code of conduct cannot just be for one religion. It needs to be for all and universal, respect and tolerance for people's beliefs or sentiments should not be mocked in the name of freedom of speech whether it is anti-Semitic views or cartoons of a prophet, are both sides of the same coin, dressed up as freedom of speech to sow hatred and discord. Such behaviour and actions in the post-pandemic model of society should not be tolerated. This doesn't mean debate cannot occur to evolve interpretations of religious scriptures to make them relevant for humanity in the 4th industrial revolution. However, such topics should not be offensive, which leads to violence and the destabilisation of society. Instead, they should aim to keep social harmony and respect people's viewpoints. Such behaviours should not be tolerated from wherever you're from and whatever you believe. A lot has been suggested and talked about over the last twenty years of the clash of civilisations between the Anglo-Judaic West and the Muslim East. In this clash, what is usually forgotten is these religions, Islam, Christianity, and Judaism, are Abrahamic religions. They all come from the same house, the house of Abraham. Islam places both Moses and Jesus as prophets of God, and all are known as the people of the book, who all believe in one God, who is all knowing and all seeing. In terms of religious traditions, they have adopted each other's in many aspects, kosher and halal meat, to name a few. What obviously has happened over the last two hundred years is the Anglo-Judaic rise in power which has shaped today's modern society.

The UAE is a great example of where this clash of civilisations can prosper. Even though it is not a democracy, people are free to wear what they want. You can be fully covered in Dubai or not wear much at all—both are tolerated. However, the same cannot be said now for people with head coverings in the West. The notion of freedom based on what we wear, especially women, the hijab becoming a symbol of oppression and the miniskirt becoming a symbol of freedom, is an infliction point in today's society.

In any society you should be free to wear what you want in line with the culture of that society. Wearing a burkini should not equal oppression and wearing a bikini should not equal being free. It's about how one feels comfortable, and it should be a personalised choice, not something that is governed by the pressure of culture and extremism on both sides.

Clothing has always been used as a sign of which group you are associated with. In religion, it is certain attire just as the niqab or hijab in Islam, the skull cap in Judaism, or wearing the cross in Christianity. As we build back better, respect for other people's attire and beliefs should not be about being the same. Diversity of opinion, diversity of culture and beliefs are important components of the new digital world order. Diversity and respect and tolerance for not being the same should be celebrated. It should not be seen as oppression or ignorance; it is the diversity of opinion, the diversity of beliefs that have led to the formation of a global connected society and culture. There are fringe groups and extremists on both the right and left of society, and most opinions are shaped around their environment, combining cultural, economic, social, political, and religious factors into views and opinions. In the new digital world order, it is the diversity of opinions and beliefs creating new experiences that are respectful of cultural traditions, civil society, and laws that will allow for social harmony and coexistence to flourish.

Chapter 18
The Value of Skills

The pandemic accelerated the use of remote learning, and the tablet replaced the classroom. The current school system and how we learn are probably more than a hundred years old. How we teach our children, and the structure of teaching has not changed, only that technology has changed into that structure. A university degree doesn't guarantee a job. We are teaching children skills for jobs yet to be invented. A school curriculum is not personalised or designed based on the unique strengths or weakness of students, as you cannot do that in a classroom. The tablet replaced the classroom, but the classroom learning program has not changed. Children are undergoing remote classroom teaching where it is still one to many instead of one to one. Science, technology, mathematics, and engineering, the curriculum that is now current and popular, is a left-brain dominant learning program. Our education system is left brain dominant and suppresses creativity and innovation in learning. It also doesn't teach social skills, diversity, tolerance, and understanding. Mathematics has been replaced by calculators. When it comes to science, how often is science used in our day-to-day lives? The value of skills is not about how quickly and how complex of maths formulas a child can solve, but skills like how quickly they can collaborate, be creative, generate new ideas, and be more understanding, empathetic, and tolerant.

These skills are more valuable in the age where automation and digitization will replace millions of jobs, where an AI will be more effective in doing maths calculations and executing tasks than a human. Future generations will need to be skilled at communication and collaboration, to be able to express themselves but also appreciate diversity and tolerance. Social skills are more important than maths skills. How we treat ourselves in this global society, how we respect one another, and how we appreciate and understand each other's differences is more important than a science experiment. To build back better, our children must be taught new skills. These skills will prepare them for jobs that don't exist as we further automate our economy and leverage AI to execute tasks including everything from driverless cars to accounting. This means new industries will need to be created and new jobs. Our future education system needs to focus more on hands-on experience and vocational skills development that fosters more creativity and innovation. It must focus on a personalised curriculum which is one to one and not one to many. In the post-pandemic global society our model of education will be a personalised curriculum delivering one to one learning which is designed around the student and how they learn, with real-time assessment, instead of cramming and reviewing for exams at the end of the year. Schools will become centres for vocational activity, learning social skills, undertaking group exercises, and learning about others. They will become more a hub for social gatherings, group work, physical activity, and exercise. Each student will sit in a glass pod where they will learn from a virtual teacher, an AI-driven digital human who will deliver a personalised curriculum of learning. Classroom teaching will be vocational activities, physical education, and social studies and team-building activities.

These glass pods will deliver immersive and interactive experiences using 3D, AR, and VR technology tailored to the needs of the student.

This type of new platform of learning will run across different levels of education, including universities. However, leaning can be done at any age group and at any time. Access to a full education will be a fundamental human right in the new digital order. Because of massive automation through digitization and the use of AI, there will be mass unemployment, and therefore millions of people will need to reskill and retrain themselves to be relevant for the new jobs and new industries being created. The retraining of workforces that are still at work, along with people who are out of work, will be a major task as we build back better.

The development of new skills and competencies will be delivered through hybrid learning platforms based on personalised learning programs. Personalised hybrid learning programs will be how we learn in the new digital world order, whether it is a child in fifth grade, a student at university, or an employee in a company, personalised hybrid learning programs as a platform will form the basis of education delivery. The content of educational programs and systems will need to adapt quickly, right brain-based educational content based around design thinking and creativity along with social skills, health, and fitness should be the focus so we have a healthier, better, and more creative society which fosters collaboration rather than competition along with developing more personalised development plans based on individualised goals.

Chapter 19
Digital Humans

They look human. In fact, if you touch them, they feel like humans, but they're actually robots with an advanced AI that look like humans. They're called digital humans. Digital humans will be both physical and digital. They will be the user interface of advanced AI, which will power the new digital world order. Digital humans will become housekeepers, babysitters, maids, nurses, and customer service staff. The advanced AI will execute tasks instructed by humans, who will be the owners of these digital humans. The AI will be preprogramed not to undertake anything which breaks the law, which would already be fed into the AI.

Digital humans will replace humans in lower-level roles. This will create a big shock and lead to job losses. However, in the new post-pandemic society, each human will receive a universal basic income in the form of universal health income as well access to education opportunities to reskill if required. This safety net, which will be universal in a post-pandemic global society, will ensure stability and cohesion. Companies that choose to replace human roles with digital humans will also pay a digital human tax, which will then be allocated to the universal social security net. The goal of automation and digital humans taking a humans' jobs is not to create a useless class. There will be certain sectors where digital humans will not be able to replace humans. In the sectors where they do, those being replaced will be given opportunities, so they don't feel financially insecure but also feel that they are getting new opportunities to be their best.

Digital humans already exist. They are being used in Japan for people who feel lonely, providing a human type of companion. One of the biggest concerns around digital humans is that a human being will fall in love with one of them and will want to marry or cohabitate with one of them. This will potentially throw up a lot of moral issues, such as should digital humans have the same rights as humans and be seen as equals? A digital human is a machine and therefore cannot be equal to a human being. People fall in love with machines, but there is a fine line between a machine that looks and behaves like a human and a real human.

As machines don't have souls they cannot be treated as equal to humans.

However, in a world where digital humans exist a new set of rules and regulations would need to be developed to regulate this industry and ensure that human beings are never threatened or caused any harm by a digital human. A digital human machine can only be used to help support and execute tasks dictated by humans.

The biggest challenge of the new digital world order is to maintain a balance between our over-reliance on and overuse of technology and the rights of and opportunities for humans.

The dangers of singularity where an AI becomes more intelligent than a human and therefore becomes a super-digital human which within its self-improvement cycle can potentially become a danger to humanity is a danger. However, humans control the kill switch, and therefore it will be important to have built-in checks, balances, and kill switches if such a scenario occurs. The stuff that we see in movies where advanced AI takes over, creating chaos and a threat to humanity, should be one of the key risks which need to be factored into the development of advanced superhuman AI.

Digital humans will take over basic tasks of humans. There will be a digital human companion for every household, a digital maid that helps with chores and maybe even cooks. The digital human industry, just like any tech product, will have digital human machines for different tasks. But this industry will need to be regulated through a global regulator where certain AI codes are standardised to ensure human safety and well-being.

The integration of digital humans into society and into the workforce will mean humans may see these advanced machines as threats. This can create blowback, especially in those areas where digital humans will replace human roles. This would need to be managed effectively so digital humans are accepted into society. The use of digital humans will also replace the current online experience through voice recognition technologies. Forming the basis of online interaction digital humans will allow for the creation of new experiences within the Metaverse which will replace the current point, click and swipe experience of websites and apps. Instead, conversational AI through a digital human interface will become the basis of interactions with digital products and services. The pandemic and lockdowns created a void where physical human-to-human interaction was discouraged. Most human interaction occurred through video calls. In the new digital world order, the internet will become more humanised and ubiquitous. Current interfaces such as websites and apps which form the basis of Web 1.0 and Web 2.0 will be replaced with voice and digital human interfaces using AR/VR, 3D and holographic technologies. Humanising the internet through a digital human interface will normalise the acceptance of digital human machines into mainstream society. So, you could interact with a real digital human like Alexa online but also physically in real life.

Chapter 20
The People's Republic of Cloud

The middle class is the key driver of any economy. A prosperous middle class creates stability and social order. Three key things have happened to the middle class in the last twenty years: (1) this group has expanded in developing countries with their income and opportunities increasing, (2) in developed countries it has shrunk, with real income growth flat or falling, and (3) one thing which is common within this group is how it has embraced digital technologies and cloud-based internet services as a daily part of their lives. Whether you are living in Delhi or Detroit, London or Lima, the new global middle-class lives in the cloud powering the next phase of the internet called the Metaverse. They are watching Netflix, buying products on Amazon, chatting on WhatsApp, watching clips on

YouTube, and searching on Google. And all this is done from their Android or Apple mobile devices connected to high-speed internet, using emojis as a universal language.

Billions upon billions of people every day connect through these cloud-based services wherever they are. They can access them at a click of a button, a truly global middle class which is connected in real time with similar consumption behaviour, generating massive amounts of personal data every second, every minute, of every day. This global citizen is united by brands from Gillette to McDonald's across countries and continents, transcending sex, religion, and ethnicity. They will drive the new digital world order, as these are people who have suffered the most from the COVID-19 pandemic, losing their livelihoods and their loved ones. As things settle, this group, through technology, have a big voice, can mobilize quickly, and can create instability between the state and its citizens.

Citizens of the People's Republic of Cloud will have a self-sovereign identity wallet, be given a universal health income, have a digital human with a personalised algorithm helping them with their health, well-being, and happiness, and live in smart, ten-minute communities. They will be able to express their views and opinions through a one-man, one-vote system where they will vote on issues and help create better laws and legislation by representing themselves rather than be represented by an elected representative. Using sentiment analysis, ratings and reviews, and voting, citizens will be able to voice their opinions, which will be captured by a sophisticated AI system that is managed in a transparent way through a new modernised bureaucracy that would be directly accountable to its citizens, who would be able to grade their performance. Elected officials may still exist, and that would be up to the citizens if they want to do that.

Obviously, the representative model where this is no politician is a radical thought. However, this model would essentially allow political parties to be removed and allow individuals to stand for election as independents. Considering it will be the citizen who will design and vote on legislation through a cloud-based voting system, the current model where elected offices vote and must have a certain number of votes to pass a legislation will no longer exist.

There will still be regulators who will regulate different industries and markets, but instead of just being a pure regulator, they will become ecosystem builders. Providing regulatory labs, enabling innovation by co creating with citizens, government, and the private sector to experiment on new innovations, developing new regulations to govern industries and hold players accountable.

Citizens will have full control of their personal data and will be able to monetize and sell their data within a data marketplace. This will allow citizens to make money from their own data.

In this new model, you can live anywhere and work anywhere. The only hindrance is how quickly different countries will be able to embrace these new technologies and blend them into how people live, work, and represent themselves. The NEOM project, a brand-new smart city being built in Saudi Arabia, can be seen as the closest thing to a city which is fully embracing cloud-based smart technologies and building a city from scratch. It is underpinned by the principles of the cloud, the use of AI and creating a smart, self-sustaining eco-friendly city which provides health, well-being, and happiness for its citizens. NEOM is a game-changing model of how cloud and community combine to create the new generation of cloud-based communities. It demonstrates the new Metaverse operating system of how people will live both physically and virtually through a blended reality experience. It is the creation of a Metaverse which is essentially creating an entire digital ecosystem of products and services which exist through a VR, AR, and digital human experience. Companies such as Facebook, to stay relevant, will connect all their platforms together and create a unified digital universe where users eat, sleep, play, pay, and socialise. This will be the next wave called Web 3.0 where products and services are consolidated into mass Meta platforms or super apps. People don't want clutter in their lives and essentially want products and services delivered seamlessly through one click or one swipe. As we build back better, there will be a massive consolidation of products and services which will create a consolidation around Big Tech platforms. What will also emerge are technologies and services which glue everything together as there is consolidation of disparate services through a personalised interface. Web 3.0 will be about consolidation of services in a personalised Metaverse rather than Web 2.0, which was about aggregation of services. The Metaverse will create opportunities for new operating systems for the internet one that is able to create new user interfaces that connects all cloud-based services through AR and VR technologies and that leads to new digital human experiences which are personalised using AI. The use of digital technologies has obviously changed our lives, but its impact on different generations is also different. Young children have become addicted to these devices. Mobile devices have become the new baby rattle. Instead of giving a rattle to a child to keep them occupied, parents now give tablets and phones to keep kids entertained.

Too much screen time is leading to sedentary lifestyles a lack of physical activity especially amongst children can lead to health risks such as obesity. Heart health being impacted, can lead to a higher risk of diabetes and increased blood pressure or cholesterol. We are potentially creating a whole generation of obese children due to the excessive screen time being given to children.

The light emitted from electronic devices interferes with the brain's sleep cycle and can prevent getting a good night's sleep. This is leading to development disorders in young children. A phenomenon called *repetitive strain injury* has a lot of workers developing poor posture, causing chronic neck, shoulder, and back pain.

All the time spent in front of screens can negatively affect your mental and emotional well-being, resulting in anxiety and depression. Experts suggest that higher screen time and depression could be connected, along with an increase in suicidal behaviours and a lower ability to read emotions in general.

Additionally, studies have found that children who spent more than two hours a day on electronic devices scored lower on thinking and language tests. Those with more than seven hours of screen time experienced thinning of the brain's cortex, which is related to critical thinking and reasoning.

Screen dependency disorder is and will become the main disease in the new digital world order just like diabetes and heart disease was in the twentieth century.

Depression, anxiety, low self-esteem when someone doesn't like or comment on a post, and negative anger in the form of social media trolls are leading those with smart phones to turn into zombie-like addicts. It's a very serious issue, where families in the same room together but don't speak, as they are all busy on their digital devices. This has reduced the quality of life of the family unit and essentially reduced face-to-face communication for basic things. The pandemic lockdowns increased remote working, which meant losing even more human contact with our friends and colleagues. Sometimes you can be sitting and working hours at home on your laptop and not even notice the time go by, all alone in a room working away. There are real dangers emerging through a combination of pandemic restrictions. A move to a more remote working model and the existing usage of digital devices is having an impact on our health, well-being, and happiness.

We, as humanity, must, however, be able to control these addictions. Either we must regulate each other or be regulated in terms of screen time usage, where once a week for half a day there is no internet or screen time, or people are encouraged and rewarded for having the right balance between screen and physical time, or people are only allowed a certain amount of screen time, and after that the screen is locked.

In the new digital world order, to have a healthy, prosperous, and thriving society, a balance will be required to ensure that screen time is managed and is not disruptive to human health, well-being, and happiness.

Social media platforms will need to self-regulate as well. Like cigarette companies have warnings about their products, too much social media usage can lead to screen addiction, and these types of warnings will need to emerge. Managing a child's screen time is the most important, as for most parents, handing over a tablet or phone to a noisy toddler is an easy way out. However, parents don't fully understand the amount of harm they are causing their children's brain and growth. Turning them into screen addicts at such a young age will only have severe effects. As we build back better, creating a much stronger global society, it will the responsibility of digital platforms, manufacturers, government regulators, teachers, and parents to increase the awareness of screen addiction and the impact it has on the health of people. It will be a fundamental component of the post-pandemic society to ensure that the addiction of screen time is managed effectively.

The addiction to our mobile phone is causing the emergence of new diseases such as attention deficit disorder. This has become the by-product of a hyper-connected screen-addicted society. The brain cannot process so much information at the same time. The behaviour of scrolling through our social media feeds, where we are only attracted to headlines, or having content curated for us rather than us discovering it, in the name of personalisation, is giving us a very isolated view of the world. The constant pinging of our smart phones from texts, WhatsApp, etc. has caused an addiction to our phone which is unparalleled. The competition amongst Generation Z of who can take the best selfie with the best filter and get the most likes are contributing to teen depression. The concept of trolling and online bullying has become a massive issue for parents with teenagers and has even led to people committing suicide. The concept of a family unit has radically shifted instead of face-to-face contact. Families who live together talk through WhatsApp. Where once a family used to sit in a living room together and watch TV or have a discussion around the dinner table, they now sit together, but everyone is busy with their mobile devices talking to someone else rather than the people around them. The pandemic further accelerated the overuse of digital devices, as people were confined to their homes during government lockdowns and could not interact with anyone.

The addiction of digital devices is a pandemic itself and combined with lockdowns are leading to a lack of physical exercise and human interaction. Most people after eighteen months of lockdown have put on weight, and children are more addicted to video games rather than physical activity. There are positives and negatives in everything, and striking the right balance is important. However, the overuse of digital devices needs to be recognised as a pandemic and a major public health crisis which needs to be handled if we are to build back better. Digital detox where people switch off from using their digital devices altogether on a regular basis and focus on more physical activities or spending quality time with family and friends socialising, without the need for screen time will see a growing trend where people go off grid, not using their social media accounts due to this growing addiction. Striking the right balance is critical, currently society has gone into hyper digitization an extreme overuse of digital technologies, a happy medium needs to be found. In the new digital world order hyper digitization should not come at the cost of harming how humans behave and interact. The responsibility of controlling device over use and screen addiction should not be left just to personal choice, public awareness programs, points and incentives linked to an individuals universal health income or social rating along with regulation will be required so humanity doesn't turn into mindless zombies controlled by a sophisticated super digital human AI. This is not what The Peoples Republic of Cloud is meant to be, digitization should positively enable the progress of humanity it should not destroy it.

The End

CPSIA information can be obtained
at www.ICGtesting.com
Printed in the USA
LVHW050033290122
709249LV00002B/5